ARTIFICIAL INTELLIGENCE MADE SIMPLE

LEARN HOW AI IS GOING TO CHANGE YOUR LIFE FOREVER

MUQBIL AHMAR

XpressPublishing

An imprint of Notion Press

XpressPublishing
An Imprint of Notion Press

Old No. 38, New No. 6
McNichols Road, Chetpet
Chennai - 600 031

First Published by Notion Press 2019
Copyright © Muqbil Ahmar 2019
All Rights Reserved.

ISBN 978-1-64783-477-7

In the memory of my father Late Mr. M.R. Khairi, who always wished to see me as a journalist and writer. Though he is no more, I hope he notices this effort of his son. For you, sir. And, of course, for you mom.

Contents

Preface

The Rise of the Machines

The Rise of the Machines

These are super exciting times! Artificial intelligence, machine learning, and robotics have finally emerged out of the pages of science fiction. They are here and now, living and breathing with us.

The writings draw on my journey as a technology evangelist and journalist over the years and someone who has been following this technology for long.

This book is for the layman as well as the expert. It tries to simplify the basic concepts of AI and introduce you to its possible impact on human beings. The chapters examine the AI technology, together with questions about business and ethics. I have also tried to cover the major debates at the center of today's tech discourse.

In the new millennium, general people, IT leaders and organizations want to know where the world is headed and what does the future look like. Will the robots and intelligent machines take away human jobs and replace humans? These questions have also been addressed along with the latest trends in the area. Look forward to your feedback. Happy Reading!

~~~~~~~~~~$$$$$$$$$$$~~~~~~~~~~~

Acknowledgements

I would like to thank my family members and the several technology leaders who gave me their valuable time for discussing the subtleties and nuances of today's tech landscape. I would also like to thank the growing frontiers of technology such as Artificial Intelligence and Machine Learning, without which this book would not have been possible. In a few hours, I was able to publish. This in itself is something that was unimaginable till about a few years back. Thanks to Notion Press for providing this excellent platform to budding writers and authors who want to share information, feelings, and emotions. Well done!

CHAPTER ONE

How AI Will Change Everything in Human Society

The Irresistible Rise of Artificial Intelligence

Technology is changing the way we live. It is also changing the definition of what it means to be a human. From artificial intelligence (AI) to cryptocurrency and e-commerce, the entire landscape is so rapidly changing that we cannot even imagine. In this context, humans, businesses and industries will have to evolve and adapt to the new realities. Some agencies like Gartner forecast and inspect how technology is changing what it means to be human. As the digital age advances, expectations around the fixed nature of what defines humans is actually starting to get challenged.

What is the first thing which comes to your mind when somebody mentions Artificial Intelligence (AI)? Many of us envision an army of human-like robots ready to take over the world!

Some may have a positive view and visualize a bright future in which it helps them in all possible ways (takes your dog out for walks or helps you get up early morning.

Well, the good news is that the truth is somewhere in between. This book will simplify the basic concepts of AI and introduce you to its possible impact on human beings.

Anyway, these are super exciting times! Artificial intelligence, machine learning, and robotics have finally emerged out of the pages of science fiction. They are here and now, living and breathing with us.

A lot of us may not have noticed but it would not take much time for them to get noticed in a significant way. In fact, the AI technology has, of late, been hogging the limelight like never before. A great amount of attention and analysis is being done to understand its functioning and how it will affect the human race.

The interest of the scientific and technical person is understandable. But even the layman is interested. The person on the street now wants to understand what the hype is all about and how it has the potential to affect his life. Needless to say, this new technology will change and influence his life in multiple and irreversible ways.

Being the latest hype in the hot and happening field of technology, everybody wants to hop on to the bandwagon. It is exciting and jazzy. Moreover, AI has also been a serious academic discipline since decades. Several of the world's leading minds have been devoting themselves to the pursuit of machine intelligence and have handled with the proposition of what it means to succeed in this pursuit.

This book therefore tries to examine the technology, business or ethics of AI as well as the topics and debates which are at the center of today's tech discourse. As 2020 approaches, general people, IT leaders and organizations are trying to read the writing on the wall and thinking about where the world is headed.

This book will also try to address the concerns and apprehension about the emergence of robots. The book tries to throw open a few very vital questions:

- *What will the future of the human race look like?*
- *Can human beings compete with machines?*
- *Will machines replace humans?*
- *And if they do what will be the extent?*

Much of the public discourse around AI has been anticipated and is being influenced by thought leaders. The book will describe what the new tech is and whether new machines can be built which have it?

What does that mean for the society? This writing effort seeks to play a significant role in raising awareness about the developments in the field of AI and its impending impact on the society.

The book is also aimed at anyone seeking a simple understanding of AI's complexities, challenges, and possibilities. For them, this can be an enlightening read as few books try to explain the consequences of the rise of the machines.

Technology in addition to its various applications are set to shake all aspects of humankind, and the environments in which homo sapiens live. Today, Artificial Intelligence (AI) is quite a common theme discussed among common people as well as in technology and business circles.

Many experts and analysts feel AI or machine learning (ML) is the future of tech. However, you look around yourself and you will find that it is not the future. It is the present.

With advancements in technology, we are by now connected with AI in some way or the other: whether it is Alexa, Siri, or Watson. Yes, though the tech is in its early phase corporations are investing funds, paving the way for a robust growth in AI-based products as well as Applications in the future.

Consider the following that would give you a sense of growth in this area!

- Between the years 2018 and 2019, organizations deploying (AI) grew from 4 to 14 per cent, according to a Gartner Survey.
- By the year 2020, more than 6 billion connected devices will be active. (Gartner)
- By 2020, "digital assistants" will use face and voice recognition across the world.
- Artificial intelligence will replace 16% of American jobs by the end of this decade (Forrester)

- As many as 15 per cent of Apple phone users use Siri's voice recognition capabilities. (BGR)
- In 2019, more than $400 million was invested in AI linked startups.

This shows an increase of 300 per cent over the previous year, says the research carried out in this area. For example, Nest is one of the most successful AI startups (acquired by the tech giant Google in the year 2014 for about $3.2 billion). It uses behavioral algorithms in order to save energy basis your schedule and behavior.

Inescapable AI: By 2025, at least 90% of new enterprise apps will embed AI, say the latest predictions from IDC. The report adds that by 2024, over 50% of user interface interactions will use AI-enabled computer vision, natural language processing, speech, and AR/VR.

~~~~~~~~~~$$$$$$$$$$~~~~~~~~~~

CHAPTER TWO

Data Analytics is the Backbone of AI

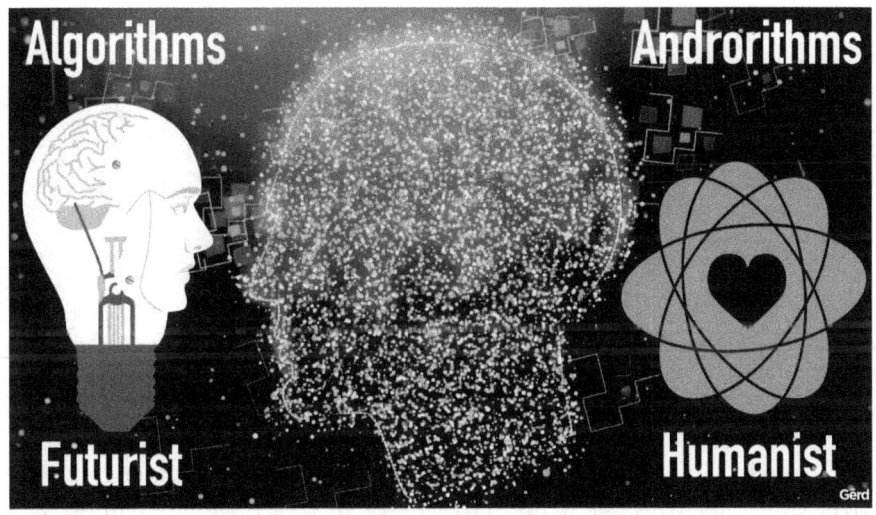

The Evolution of Future Machines

Artificial intelligence (AI) is the simulation of human intelligence by machines, especially computers. It focuses on the development of intelligent machines, which think and work like humans, for instance, speech recognition, learning and planning, and problem-solving.

The tech also includes reasoning (using guidelines to reach conclusions) and learning (acquisition of information as well as rules for using the information), in addition to self-correction. Particular applications of AI include expert systems, speech recognition and machine vision.

Data Analytics is the Backbone of Artificial Intelligence

AI is primarily a combination of various technologies. In this, machine learning is one of the main techniques that is used. This combination of AI and machine learning makes various assumptions, re-evaluates models and data. All this happens without the intervention of a human being. A human engineer does not need to code for every possible action. AI is able to test and re-test data in order to predict all possible matches, at a speed and capability no human could attain.

AI and Machine Learning algorithms learn and provide predictions at a scale and depth of detail that is impossible manually. Complex analysis can be done in real time with many variables involved. This allows the system to learn rapidly, which in turn, can deliver micro-target insights which cannot be done by human analysts on a large sample size of population. The results can improve conversion rates, for example, marketing returns on investment as well as customer loyalty. Therefore, it is imperative that enterprises, governments and marketers comprehend benefits of this new technology.

Significance of Data Analytics

Data mining is a process that is used to deliver big quantities of data, which is often unstructured. The data has to be analysed and made sense of; however, this is impossible till the data is given a proper structure. Those more familiar interacting with the data through dashboards understand that structured data is needed in order to achieve analysis of commonalities, like averages, ratios, or percentages. The objective is to collect data to search for a pattern or find relationships between different variables. At best, human beings can make assumptions by analysing that data.

The data has to be queried in order to attest to a certain relationship or pattern. If found valid, such a testing can continue on additional data. Data analytics of this kind is largely descriptive as it is based on past happenings. Therefore, it can't predict the impact of a change in some variables. Enterprises generally have the ability to engage with as well as process data analytics, due to their access to mountains of data which are at their fingertips, including user data on apps as well as websites, lead conversion rates, online click-through, and CRM data analysis.

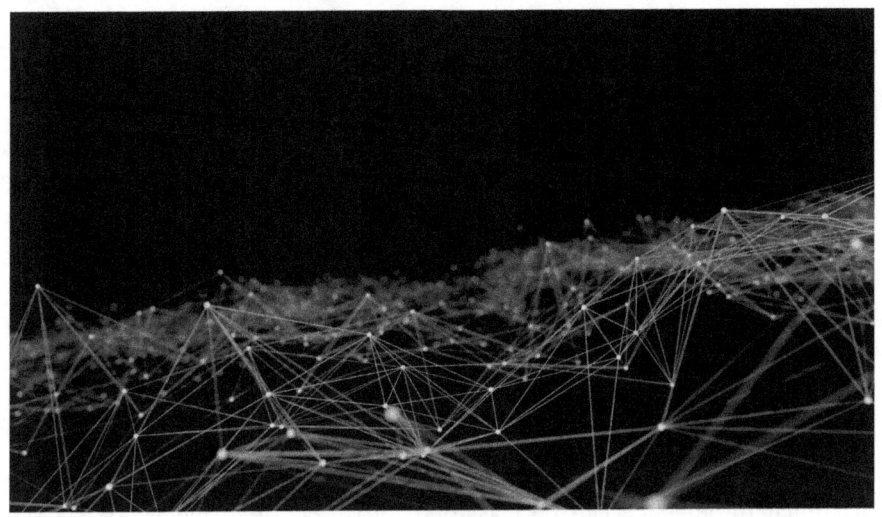

Predictive Analytics

Predictive Analytics

This is where this new term comes into the picture. Data analytics can lead to predictive analytics using collected data to predict what might happen next or in the future. Predictions, which are based on historical data and rely on human interaction to query data, can authenticate patterns and create and test assumptions. Assumptions drawn from the various past experiences generally pre-suppose that the future would follow roughly the same patterns. Such What/if assumptions can be done through a human understanding of the past, together with a predictive capability that is limited by volume, time, as well as cost constraints of human data analysts.

Predictive insights can be derived from analytics and can be very handy for companies and governments. They can help predict the effectiveness of decisions and can lead to informed decision-making on various variables, such as collateral, geographic, or demographics. A highly detailed segment can be targeted at a reasonable time and cost. Otherwise a hyper-personalized campaigning can be almost impossible. It is for this reason that AI tools that impact marketing are growing rapidly and are estimated to cross $40 billion by the year 2025. Most companies are aware of this potential of AI.

However, a number of them are unsure or unaware of the extent of benefits that can be drawn and how AI can be adapted to suit their individual requirements to improve their efforts such as marketing. AI and Machine learning is a continuation of predictive analytics, with one main difference. AI can make assumptions, test and learn autonomously.

These advances mean that developers can create innovative as well as cutting-edge products and services which would not have been within the reach of small companies. The new products as well as services which enter the market make AI adoption at a lower risk with focus on delivering practical as well as impactful results.

What is Weak and Strong AI?

Artificial Intelligence can be classified into weak or strong AI. Weak AI, which is also known called narrow AI, is designed for a specific task and is restricted by purpose. Apple's Siri or other virtual personal assistants are considered as forms of weak AI.

Strong AI, which is also termed artificial general intelligence, is empowered with generalized human thinking abilities.

The difference between both is that a strong AI can find a solution without any human intervention, particularly when it is confronted with an unfamiliar task. Deep Blue and Google's AlphaGO were both designed for restricted purposes and therefore they cannot be applied to other situations.

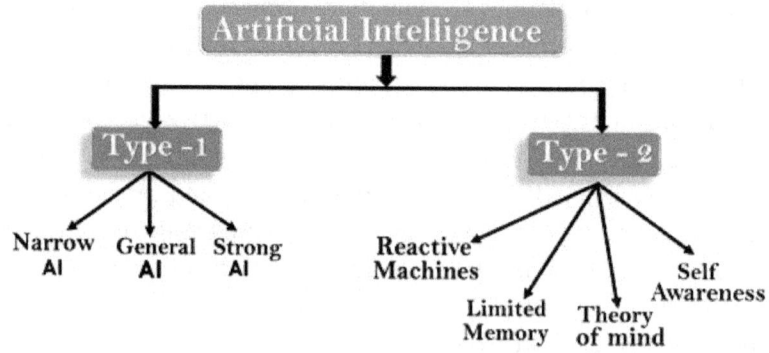

Types of Artificial Intelligence (AI)

A professor of computer science and engineering from the Michigan State University, says that the AI can be categorized further into 4 broad categories.

- **Reactive machines:** An example is Deep Blue, the IBM chess program which defeated chess champion Garry Kasparov in the 1990s. The AI identified pieces on the chess board and made predictions. However, it had no memory and therefore was not able to use past experiences to influence future ones. It anticipated moves, both its own as well as its opponent's. Based on observations, it opted for the most tactical move.
- **Limited memory:** These AI technologies can leverage past experiences in order to make informed decisions in the future. Such decision-making capabilities are embedded in self-driving cars. Observations decide actions in the near future, such as when a car is changing lanes. However, the observations don't get stored on a permanent basis.
- **Theory of mind:**This term from psychology refers to an understanding that other systems can exist which have their own set of beliefs, desires, as well as intentions and which can have an impact on the decisions that they make. Such type of AI system has not yet been developed.

Self-awareness: In this category are included systems which have a sense of the self or in other words they have consciousness. These machines are empowered with self-awareness to understand their current state and use that information to infer what others are feeling. This type of AI technology also has not yet been developed.

~~~~~~~~~$$$$$$$$$$$~~~~~~~~~~

CHAPTER THREE

Examples of AI from Daily Life: Apple's Siri

We all know how conversational AI has gained great popularity among people due to the worldwide success of factors such as Amazon Alexa, Google Assistant and others. The following five intelligent AI solutions are being used every day as we speak and changing our lives in many ways.

Say Welcome to Siri

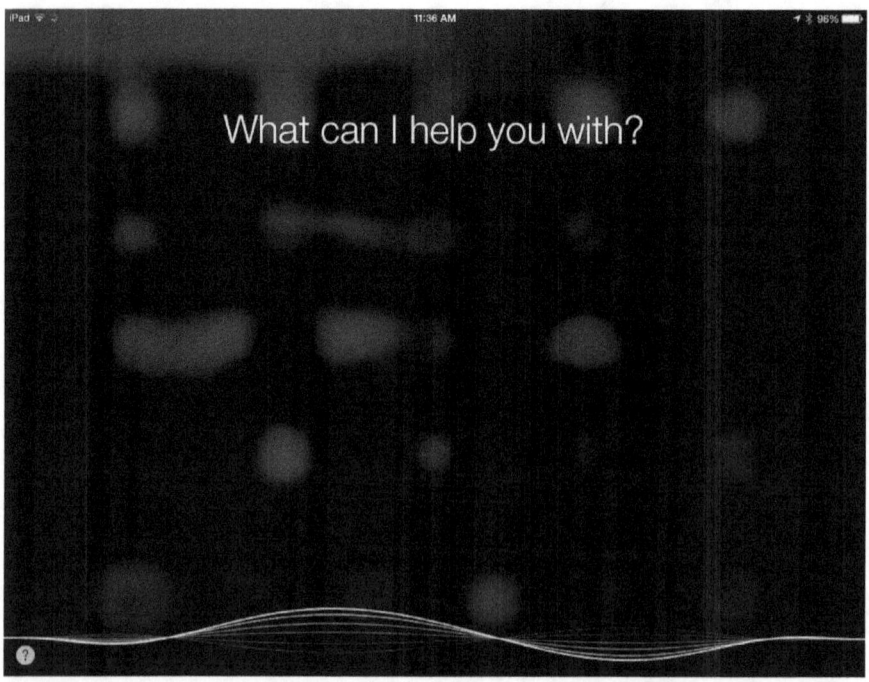

Caption: Apple's Siri's Welcome Screen

These are some of the most popular personal assistants in use today. The friendly voice-activated assistant communicates with users on a daily basis. It helps people get information, directions and messages, make calls, open applications and manage your calendar.

Siri utilizes the machine learning (ML) technology to get smarter and is able to understand natural language questions as well as requests. It is one of the most famous example of gadgets with in-built self-learning.

In fact, virtual assistants like Siri are making millions of voice calls demanded by humans on a daily basis. This tool can analyze the human voice to provide guidance in real time. It can also help in tricky situations and enhance your ability to react to a context and bring about suitable changes in behavior.

~~~~~~~~~~$$$$$$$$$$~~~~~~~~~~

CHAPTER FOUR

Tesla: Your Intelligent and Autonomous Car

Caption: Tesla Roadster

Everybody knows about Tesla, particularly so if you are a tech geek. And if you are particularly inspired by cars shown in Hollywood flicks, Tesla is what you want in your garage. The company makes the most technologically advanced automobile in the world and is a pioneer in building autonomous vehicles, incorporated with AI.

So, it is not only smartphones which can boast of AI but also cars. The vehicles get smarter every day due to updates over the Internet.

The car incorporates features unheard of before such as self-driving, predictive abilities, and state-of-the-art technological innovation. Tesla's autopilot is a technologically advanced system for driver-assistance feature which has capabilities like adaptive cruise control, lane centering, self-parking, autonomous navigation on limited access freeways, and automatic lane change.

The company has a huge number of sensors gathering data on the roads than anybody can possibly think of. It has also ensured that the technology doesn't remain the preserve of a few and has launched its first mass-market car, the Model 3.

The AI engine of Tesla is run by the data from the cars and their drivers. Internal and external sensors pick up every information such as a driver's hand placement on the instrument cluster and how they use them. Tesla cars send that data directly to the cloud. This helps Tesla refine its systems and gives it the ability to resolve issues remotely.

A famous example is the resolution of a heating problem within the car's engines. Data was showing that the components of the car were overheating during certain times. With the help of just a software patch, every car was automatically repaired.

While most of the above happens in the cloud, it is edge computing that helps the car take action at an individual level. The new technology – driven by a partnership with hardware manufacturer Nvidia – is largely based on an unsupervised learning model of machine learning. **(You can also summon the car to and fro a garage or a parking spot.)**

Interiors of Tesla Model 3

The data generates highly dense data maps which can display a number of aspects such as average increase in traffic speed and identification of hazards for which drivers need to take action.

There is another level of decision-making as well. The cars are capable of forming networks with other Tesla vehicles close by and share local information and insights.

In the future, when autonomous cars become common, the networks may also include interactions with cars from other manufacturers and further systems such as mobile phones, traffic cameras, and road-based sensors. Plus, the data in itself is incredibly valuable. McKinsey and Co researchers have valued the vehicle-gathered data at $750 billion a year by 2030.

~~~~~~~~~~$$$$$$$$$$~~~~~~~~~~

CHAPTER FIVE

AI-Powered Netflix Ushers In Next-Gen Entertainment

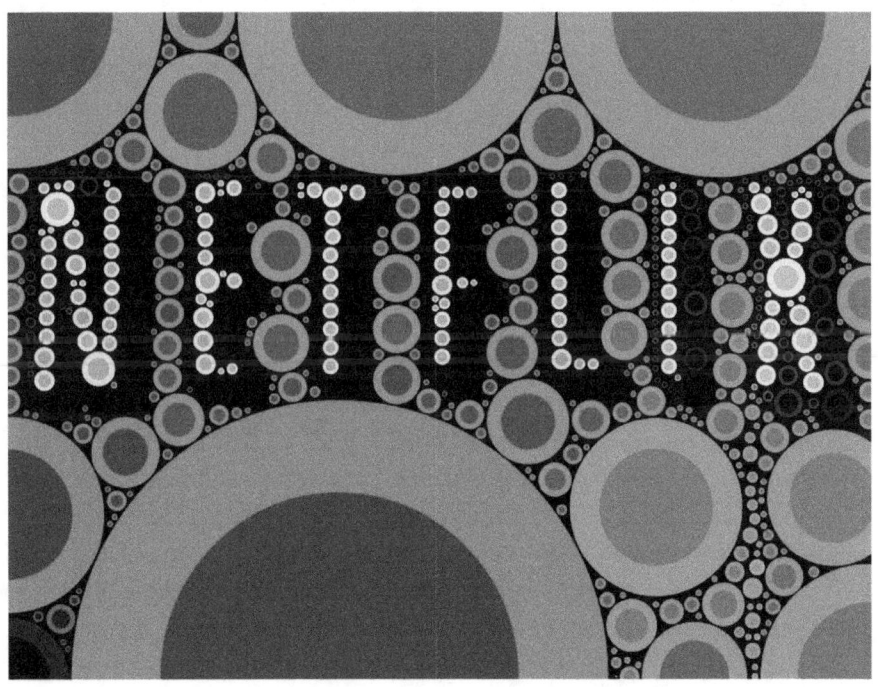

Netflix Revolutionising Entertainment

Who doesn't use or know about Netflix? It doesn't need any introduction. The video on demand platform has revolutionized the way we consume entertainment. The presence of Artificial Intelligence (AI) is becoming ubiquitous. Big corporations such as Netflix, Facebook, Amazon,

Spotify, etc. are using AI tools which interact with customers every day. These have the ability to improve over time and provide scale, significantly helping both business as well as users.

The hugely popular content-on-demand video service utilizes predictive technology in order to offer recommendations on the basis of consumers' interests, reactions, choices, and/or behavior. It uses a combination of machine learning and behavioral science for enhanced customer experience.

The AI engine makes movie recommendations on the basis of analyses of your previous choices and reactions. With the gathering of huge amounts of data, the algorithm is getting increasingly more intelligent with every passing year. The new-age company has done a remarkable job of applying AI, data science, and machine learning in the right mix.

Netlix's product-based approach is particularly good for business. The approach focuses on a business need first, then an AI solution next, instead of the other way round. Netflix has demonstrated that applied properly, AI can do wonders.

How Netflix Uses AI/Data/ML

1. **Personalized Movie Recommendations:** A user watching A is likely to watch B as well. Everybody knows about this feature. It also leverages the watching history of different other users who have similar tastes in order to recommend what you could be most interested in watching next. This strategy helps it keep you engaged and, in turn, persuade you to continue your monthly subscription for more time.

2. **Auto-Generation of Thumbnails or Artwork:** This feature utilizes video frames from a movie or show which become starting points for thumbnail generation. The company's software uses the images in order to rank every image to identify thumbnails that have the greatest likelihood of resulting in a click by you. The calculations are on the basis of what other people similar to you have clicked on.

3. **Location Scouting (Movie Production):** Utilizing the data, the location and timing of shoot can be decided on a movie set. Given the limitations of scheduling (actors and crew member availability), budgets (venues, flights/hotel costs), as well as production scene requirements (day or night shoot, the risks of weather-related events at a location), the data

is helping in such scenarios as well. It is an optimization problem whose solution is helped by a machine learning model which makes predictions based on the past data.

4. **Editing of Movies:** Throughthe use of historical data of quality checks which have failed in the past (syncing of subtitles to sound and movements) to predict when a manual check is most beneficial. This can be a tedious and time-consuming process otherwise.

5. **Quality of Streaming:** The AI engine uses past viewing data in order to predict bandwidth usage to help Netflix decide when to cache the regional servers. This helps in faster loading times during the peak of demand.

The applications of data science or machine learning at Netflix have ensured a scalable impact that has forever changed the landscape of technology as well as user experience for millions of people around the world. The leveraging of AI-related tools will only get strengthened over a period of time. However, before the astounding success of such solutions, Netflix connected the AI tools with a business need. They created a business link. The use cases developed only due to connecting Netflix's core business problems with ideas.

~~~~~~~~~~$$$$$$$$$$~~~~~~~~~~

CHAPTER SIX

Flying Drones, Alexa and Echo

Alexa and Echo

Amazon Alexa

Amazon Alexa, known simply as Alexa, is a virtual assistant AI developed by Amazon. It was first used in the Amazon Echo and the Amazon Echo Dot smart speakers.

The revolutionary piece of technology, which is a great hit with the young generation, can help you search the web for information, shop, schedule appointments, control lights and switches, answer questions, read out audiobooks, report traffic and weather, and provide info on local business and activities. All this is achieved by using the Alexa Voice Service.

With each passing year, Amazon's Echo is getting smarter, with the continuous addition of new features.

Flying Drones

A self-maneuvering flying drone

If you are aware of the changing technology landscape, flying drones are already delivering products to customers' homes (although the tech is not that widely used and is currently on a testing mode).

The technology uses a machine learning system coupled with a complex algorithm which can decode the surroundings through the use of sensors and video cameras.

This is visualized as a 3-D model. An algorithm generates a trajectory which helps a drone navigate through the space. The technology also helps it decide how and where to move.

Through a Wi-Fi system, the drones can be controlled and used for particular purposes such as video-making, product delivery, or news reporting.

Conclusion

Artificial Intelligence is gaining popularity at a quicker pace than we may realize. And, it is deeply influencing the way we live and interact, particularly through consumer-oriented products.

~~~~~~~~~~$$$$$$$$$$~~~~~~~~~~

CHAPTER SEVEN

Artificial Intelligence Jobs Are the Sexiest Today

AI Jobs Are the Best in the Market

The AI spectrum is expanding from autonomous cars to cancer detection. Such is its scope that it will touch every domain of human activity. AI and machine learning will generate an extra $2.6 trillion in value for marketing and sales by the year 2020 and around $2 trillion in manufacturing and supply chain planning, according to figures the McKinsey Global Institute.

Gartner says that the business value created by AI would reach $3.9 trillion in 2022. On the other hand, IDC predicts that global expenditure on artificial intelligence systems would reach $77.6 billion during the same year.

This has created a huge demand for skilled workforce. Companies need highly-skilled AI workers to develop as well as maintain a wide range of software applications.

No wonder, AI specialists and trainers, machine learning engineers, natural language processing experts, deep learning engineers, computer vision experts, and data scientists are some of the most trending job profiles these days. However, there is a huge shortage in supply.

The number of potential employees is very less and job seekers are few. In this section, we will find out which tech jobs are the most in demand these days? What are the most sought after and well-paying AI jobs? Where is the demand strongest?

Let's have a look at some of the numbers:

- Employer demand for AI-linked roles has doubled over the last three years, says a job search portal. Its report says that the number of AI--related job postings as a share of all job postings is up by almost 119%. In the future, the demand is set to spike.

- Job seeker interest is at an all-time high (more than twice than it was three years ago). The top three most in-demand AI jobs in the market are: data scientist, software engineer, and machine learning engineer.

IT companies will humongous amounts to AI professionals to keep them in their organizations.

AI Creating Value in the Employment Market

Boost in AI Jobs

The following are some figures and data related to AI salaries, shifts in supply and demand dynamics, hiring geographies, and the total amount of jobs AI may create.

$142,858: This is the average for the highest paying AI job title of 2019. In the same region, are salaries for machine learning engineers, according to Indeed, a job-posting site. The salaries for AI professionals are up 5.8 per cent over the past year. This is quite above the average 2.9 per cent expected by human resources consultancy Mercer. The top five paying AI positions were data scientists ($126.927), computer vision engineers ($126,399), data warehouse architects ($126,008), and algorithm engineers ($109,313).

Deep Learning Engineers Second most in-Demand AI Job

What is surprising is that deep learning jobs weren't even on the list the past year. Deep learning engineers currently rank second on Indeed's list of the top roles in job listings that seek AI or machine learning skills.

There is a particularly high demand for experts of a particular domain of machine learning, in which algorithms are motivated by the structure and function of the brain. Such roles are in great demand in fields such as advanced robotics, autonomous driving, and facial recognition systems. The speed of AI job posting growth has increased over the last year.

In fact, the year on year growth in volumes of AI-related job posts was about 32 per cent in January 2017 to January 2018, compared to only 20% for the same time period in the years 2016 and 2017, the Indeed report has found.

The increase is driven by a mounting bulk of postings for machine learning engineers in addition to computer vision engineers. Deep learning may account for $3.5--5.8 trillion in annual value, says a McKinsey Global Institute report.

Top Market Areas for AI Job Listings

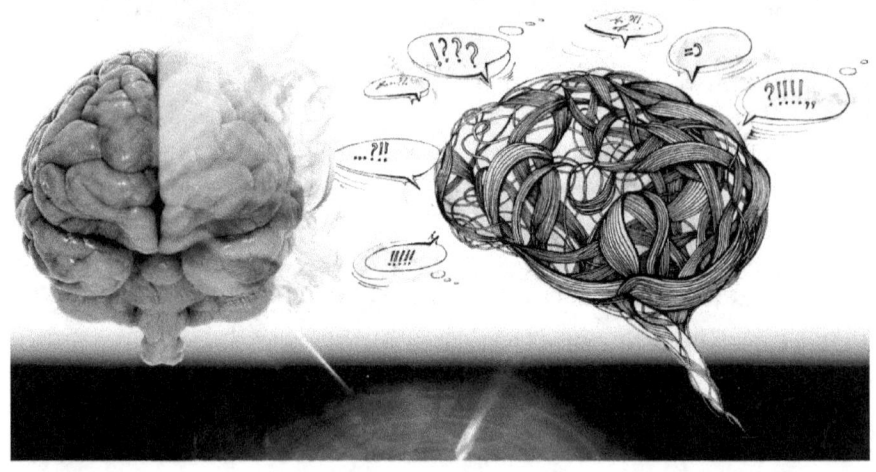

AI Effect on Job Market

There are 7,000 AI job openings in the United States, according to RPA firm UiPath. China becomes the leader with 12,000 openings, and ahead of the United Kingdom, India, Germany, France, Canada, Australia, and Poland. The top AI job markets in the USA are in New York, Washington D.C., San Francisco, San Jose, and Seattle. These top areas for AI job listings have not changed from the organization's previous analysis.

"Indeed's blog says: "New York's top position could surprise you until you consider that it is home to diverse industries, ranging from financial services to publishing and many of these are adopting AI. Also, West Coast-based tech companies (like Amazon, Facebook, and Google) have a substantial presence in this region. New York has its share of AI-based tech startups like Clarifai, AlphaSense, Persado..."

"Data shows there is growing requirement by employers for AI talent, particularly for the positions of data scientists and machine learning engineers," says Indeed economist and author of the report Daniel Culbertson. "As AI adoption continues across industries, there will be greater competition for this talent."

Amazon, Microsoft, Apple, KPMG, and Booz Allen Hamilton have topped the list of firms with the maximum share of AI postings in the USA, the study has found.

"Data scientist jobs top the list. "AI is effective when it is fueled by huge sets of quality data," Culbertson says. "Data scientists, who can analyze as well as interpret data can benefit businesses better appreciate and influence crucial understandings from the data to achieve their commercial goals more resourcefully and effectively."

While there is stiff competition across all job domains across all sectors, machine learning and AI jobs and employment opportunities face a considerably reduced amount of competition. These employment opportunities are increasing but they still remain vacant due to the shortage of skilled workforce. This is the best time for IT professionals to upgrade themselves and stay relevant.

Balancing Workforce Diversity: AI Use Case Deep Dive

Ensuring diversity in Workforce

Workplace diversity brings people across races, gender, ethnic groups, age categories, personality types, thought processes, educational levels, and backgrounds together. Diversity in workplace leads to flow of varied ideas and different approaches to the same problem. It reflects on how an individual perceives himself with respect to others, thus, encouraging an intelligent and healthy work environment.

What is Diversity in the Workforce?

Workforce diversity means creating an inclusive and accommodating environment which accepts every individual's differences, strengths, and weaknesses and also provides opportunities to all members of the staff to achieve their potential.

Valuing and prioritizing difference and diversity allows a person to contribute their own unique experiences to a workplace. This has positive impact not only on internal activities or relationships, but also on customers as well as other stakeholders. According to data, organizations which are diverse, are more innovative, smarter, and retain more staff. Moreover, diversity makes for good business in the financial sense.

Businesses, with greater diversity, witness increase in returns on investment (ROI); around 35 per cent for an ethnically diverse workforce and 15 per cent for gender-diverse organizations. Companies are able to win prime talent and enhance employee satisfaction, so workers tend to stick.

No wonder, Canadian Prime Minister Justin Trudeau appointed a gender-balanced cabinet comprising 50 per cent women and 50 per cent men in 2015. French President Emmanuel Macron also appointed a cabinet diverse in gender and political leanings.

Corporates spend large amounts of money on training in order to build a diversified workforce. However, according to latest research such volunteer-driven efforts have not borne fruit as was expected. Despite the efforts, bias continues to exist in recruitment against a diverse workplace. Point is it is difficult to eliminate unconscious bias from organizations.

Psychologists also agree that everybody holds unconscious beliefs and prejudices against various social and ethnic groups, and that this bias stems from one's tendency to organize social worlds into categories. Researchers have proved that we let extraneous factors influence our decisions.

We often end up pre-judging candidates on the basis of unrelated factors such as educational background or place of origin. According to psychologists there are three key reasons for bias related to recruitment for a diverse workforce:

Similarity inclination: Generally, people are inclined towards hiring candidates who are similar to them. According to the similarity-attraction hypothesis, one has affinity towards people of similar character and that could mean sharing the same religious background or liking the same sport.

Physical appearance: Interviewers form an opinion about the candidates on the basis of their appearance ranging from their hairstyle, tattooed body or complexion. According to a study conducted by German Researchers, people underestimate the ability of obese individuals to achieve supervisory positions as compared to 'normal-weight' individuals.

Stereotypes: Interviewers invariably go with stereotypes. If a candidate is not from a known university, he may not be considered a potential candidate, or a female candidate may not be deemed fit for a number crunching role or sectors which are technology oriented.

Organizations must find out ways to put in place a diverse workforce. This will have far reaching ramifications for the organizations own health and character and bodes well for a well-knit and caring society that is rid of discrimination and strife. The advantage of AI over humans is that it can analyze a vast amount of data and search for patterns within lesser time compared to humans. AI collects and churns out data from various sources to find people based on the job description.

Candidates who may have never applied in the organization and whom the company would have never thought of seeking out, can also be shortlisted by AI. This application can remove unconscious bias and focus on the skills required for the job ignoring other information such as a candidate's age, gender, and race.

However, AI identifies the past behavior patterns which means any human bias that may already be in the recruiting process – can be learned by AI. For example, if earlier the recruitment was done from a preferred college, AI will pick up the same pattern repeatedly.

Thus, it is critical to take precautionary measures to remove unconscious bias. AI can be taught to filter irrelevant considerations out of the decision-making process; pick the most suitable candidate based on objective criteria.

~~~~~~~~~~$$$$$$$$$$~~~~~~~~~~

CHAPTER EIGHT

Man Vs. Machine: Will Humans Lose Jobs to Computers?

Human Vs. Machine: Will People Lose Jobs?

"Sooner or later, the U.S. will face mounting job losses due to advances in automation, artificial intelligence, and robotics." —Oren Etzioni.
Tesla CEO and billionaire entrepreneur Elon Musk has also said that AI poses an "existential threat" to human civilization.

"AI is a fundamental existential risk for human civilization, and I don't think people fully appreciate that," Musk says. He explained

that he has access to really cutting-edge AI technology, and on the basis of what he's seen, AI is "the scariest problem."

Musk is the co-founder of OpenAI, a research body dedicated to ensure that AI is develops in a safe and manageable way in order to minimize existential risk robots may one day pose to humanity. "

Chinese billionaire and Alibaba founder Jack Ma too has sounded the alarm bell. He says that in 30 years, the world's best CEO will be a robot. The maverick entrepreneur, who is worth £22.8 billion ($28.4 billion), says that advances in technology would render CEOs irrelevant in the next 30 years.

According to a survey, around two out of three British people are of the opinion that an "android takeover" in the future is a "genuine concern."

There is a concern that robots and machines will become smarter with AI and take over tedious and manual tasks which could eventually threaten the stability of the society.

Job Losses Galore Due to AI and Automation

Infosys, India's second largest IT major, cut thousands of jobs due to automation. While releasing the figures during the Annual General Meeting of the company, the "Admission" of having "released" thousands of people due to implementation of automation gives the impression that this "release" was forced upon Infosys, when it is actually not true. The company has been actively working towards that goal. This is but a euphemism for planned layoffs that were always in the making. Such developments have sparked a debate over increasing job losses in India's IT sector.

Similarly, tech giant Microsoft laid off thousands of employees due to its focus on cloud computing and automation. This is not all. The same has happened at some of the other big IT companies of the world such as TCS, Wipro, Cognizant, IBM, Tech Mahindra, etc. over the past few years.

According to a report by a top HR firm, 5–6 lakh engineers would lose jobs over the next three years, that is, between 1.75 and 2 lakh per year for the next three years.

"Contrary to media reports of 56,000 IT professionals losing jobs, the actual job cuts will be between 1.75 lakh and 2 lakh per year in next three years, due to under-preparedness in adapting to newer technologies," Head Hunters India Founder-Chairman and MD K Lakshmikanth said analyzing a report by McKinsey & Company at the Nasscom India Leadership Forum.

Ground Work for the Firings Had Been Laid Year Ago

Sackings themselves are not surprising; in fact, they were always underway. Wipro laid the foundation for the layoffs way back in 2015. The IT major was expecting the headcount to decrease by $1/3^{rd}$ as the company goes ahead with automation of jobs. It is part of business strategy as Wipro pursues extensive automation, artificial intelligence, and digital services. Most of the other IT companies have been forced to follow suit to keep the business margins intact.

Machine Bosses to Take Over?

"I visualize a time when we will be to robots what dogs are to humans, and I'm rooting for the machines." —Claude Shannon.

Jack Ma's statement that that advances in technology will render CEOs irrelevant in the next 30 years seems chillingly true. The situation throws up a number of pertinent questions:

- Will robots replace humans?
- Will modern technology leave hundreds of thousands of human beings without a proper occupation?

The competition for jobs and employment has reached a new level.

Can Humans Keep Up with Machines?

Jobs Losses can Hit Society

" "There is no reason and no way that a human mind can keep up with an artificial intelligence machine by 2035." —Gray Scott. "

The rivals this time are highly-sophisticated machines and robots and they are learning fast and deciphering patterns using sophisticated algorithms as well as mathematical models. Armed with big data, artificial intelligence, machine learning, virtual reality, Internet of Things, augmented reality, such machines will outdo human beings in almost all areas. Homo sapiens can be no match for advanced machines that can crunch unimaginable quantities of data in seconds. No doubt, there will be far-reaching and irrevocable changes in the world's economy.

Why Artificial Intelligence May Not Be Such a Bad Thing for Jobs

A global survey by Allegis had these insights:

- 21 percent of the people viewed AI as something to be excited about.
- 17 percent considered it both disrupting and enabling
- 9 percent, believed AI will displace most jobs in the coming 10 years.

> *"I believe this artificial intelligence is going to be our partner. If we misuse it, it will be a risk. If we use it right, it can be our partner," says an expert.*

But the fact of the matter is that automation coupled with AI could threaten 8 million jobs by 2030. At the same time, technology is all likelihood is set to generate additional jobs, some say more than ever before. They also say that therefore AI should not be seen entirely in a negative light and it may not after all be purely destructive. They say automation requires creating new skill sets.

> *"We need to address the challenge. In the early 60s, we would reach out to the nearest branch to withdraw money but with ATMs in place, the process is gone. ATMs for cash disposal, robots for painting of cars, computers for creation of spreadsheets, did feel like a threat firstly. Comparable could be the case with AI, we never know," he adds.*

While unskilled jobs may be faced with grave threats, AI could create space for a new category of jobs which could be mastered with the help of training, says an expert from the automation sector.

> *"If you don't believe in this, this is the right to start believing in it. Though a lot of jobs may take a hit due to AI, you could still land a job which involves constant learning and up-skilling," opines the industry pundit. "New jobs could be created, whereas existing roles could be re-built. This could be a great opportunity for switching careers and embarking on a new journey professionally," adds the industry insider.*

There are many industry reports which say more than 30 per cent of jobs could be under a potential threat as people continue to lose their jobs. Others say that AI tech will create more than 2 million job opportunities globally by the end of the year 2020.

~~~~~~~~~~$$$$$$$$$$~~~~~~~~~~

CHAPTER NINE

World Economic Forum (WEF) Report on Job Losses due to AI and Automation

World Economic Forum (WEF) Report on Job Losses due to AI and Automation

The concerns about machines and robots taking over human jobs have been validated by an authentic report. Research from the World Economic Forum forecasts that by 2025, machines will perform more current work tasks than humans, compared to 71% being performed by humans today.

The report goes on to add that machines will perform more than half of all "work tasks" by 2025, compared to just 29% today, which will lead to the displacement of 75 million jobs between now and 2022. However, the study also goes on predict that over the same 5 year period, rapid evolution of machines and algorithms in the workplace could create 133 million new roles in place of 75 million that will be displaced. This will lead to 58 million net new jobs.

There will be a profound effect on the global labor force. The research, published in the **Future of Jobs Report**, is an attempt to understand the potential of new technologies such as Artificial Intelligence, Big Data, Internet of Things (IoT), etc. to disrupt and create jobs. It also seeks to provide guidance on how to improve the quality and productivity of the current work being done by humans and how to prepare people for emerging roles.

The report is based on a survey of chief human resources officers and top strategy executives from companies across 12 industries and 20 developed and emerging economies (which collectively account for 70% of global GDP), the report finds that 54% of employees of large companies would need significant re- and up-skilling in order to fully harness the growth opportunities offered by the Fourth Industrial Revolution.

At the same time, just over half of the companies surveyed said they planned to reskill only those employees that are in key roles while only one third planned to reskill at-risk workers. While 50% of all companies expect their fulltime workforce to shrink by 2022 as a result of automation, almost 40% expect to extend their workforce generally and more than a quarter expect automation to create new roles in their enterprise. The rise of intelligent machines will no doubt take away jobs but they will also create new ones and the balance will be in humanity's favor.

133 million new jobs will be created by AI by 2022: WEF Report

The report goes on to say that the above is the number of jobs created due to AI-enabled automation by the year 2022, according to the World Economic Forum's (WEF) 2018 Future of Jobs report. This twice the 75 million job roles that the WEF research says will be displaced.

Automation and AI May Lead to Creation of New Job Categories

Extreme automation and advancements in technology may turn out to be good and beneficial for humans. The predictions are in stark contrast to the general concerns that new technology such as Artificial Intelligence (AI), Big Data, etc. will take away jobs in droves and leave human beings redundant. However, it is difficult to predict if the displacement of jobs will not lead to struggle and conflict among people who would have lost their livelihood due to the digitalization of the society.

The report presents a vision of a future global workforce that provides grounds for both optimism and caution. The outlook for job creation is much more positive than before as businesses have a much greater understanding of the opportunities made available by technology.

At the same time, the huge disruption automation will bring to the global labor force is almost certain to bring with it significant shifts in the quality, location, format and permanency of roles that will require close attention from leaders in the public and private sector.

Of all the advancements, Human judgment is less likely to be surpassed by AI. We need to start looking at Artificial intelligence in its purest form- Intelligence in augmented form instead of a job-hungry robot training to take over.

Will AI Create More Jobs by 2025?

Any technology which has the prospective potential to bring about alteration in human lives has produced a colossal amount of debate. This is true for Artificial Intelligence as well. In fact, the debate seems never-ending.

There are diverging opinions on AI and its probable impact on humanity among thinkers, IT professionals, researchers, and even the average person on the street. The mixed view of AI is unsurprising as technology does more than automate tasks, it fundamentally changes the nature of the work.

As AI becomes able to carry human-like functions, it will replace jobs. There is no doubt about it. But it could create new opportunities as well, say many tech pundits. According to Gartner, the entire hype about AI replacing jobs has a lot to do with the type of jobs under consideration. While certain aspects of a job are susceptible to automation, human intervention cannot be replaced.

- As many as 40 per cent of industry experts and CEOs say that they're adding jobs due to AI.
- Around 40 per cent of respondents from Global 2000 organizations say that they are adding more jobs as a result of AI adoption, says a 2019 Dun & Bradstreet report.
- On the other hand, only 8 per cent are cutting jobs due to the new tech.
- Only 8 per cent of the 100 executives say their corporations were planning to cut jobs due to the implementation of AI capabilities.
- About one third (or 34 per cent) affirmed that there was no change in HR needs due to the coming of AI.

AI Can Power Multiple Aspects of Today's Jobs

New Avenues of Job Growth?

Machines are great at cyclically performing a particular task with a high level of accuracy and consistency, thus we can surely say AI will take over a particular variety of tasks. But complex problem solving remains a far-fetched thought among the goals of AI. Change is the only constant and as some jobs are replaced new ones will be created.

As per some estimates, 65% of the kids who are in schools today will end up with jobs which do not exist today.

When it comes to tactical thinking, nuances of problem-solving, adaptive thinking, and thinking-out-of-the-box abilities, AI is still way behind the human brain. After all, AI can only mimic the human brain.

So it seems AI will cut off the monotonous jobs on the table (like data entry and a certain level of accounting) but human resource-based jobs like customer care, sales and marketing, innovation, and research will continue to be in high demand along with specialized jobs in the field of AI itself.

For those of you who are wondering what should you do to save your job- a little less worry and more upskill and training will set the game just right for you.

Advantages of AI over Human Intelligence

Advantages of AI over Human Intelligence

"According to the father of Artificial Intelligence, John McCarthy, it is "The science and engineering of making intelligent machines, especially intelligent computer programs". "

Artificial Intelligence is a way of making a computer, a computer-controlled robot, or a software think intelligently. The advantage of AI over humans is that it can analyze a vast amount of data and search for patterns within lesser time compared to humans.

For example, in hiring and recruitment, AI collects and churns out data from various sources to find people based on the job description. Candidates who may have never applied in the organization and whom the company would have never thought of seeking out, can also be shortlisted by AI. This application can remove unconscious bias and focus on the skills required for the job ignoring other information such as a candidate's age, gender, and race.

However, there can be in-built disadvantages too. AI identifies the past behavior patterns which means any human bias that may already be in the recruiting process – can be learned by AI. For example, if earlier the recruitment was done from a preferred college, AI will pick up the same pattern repeatedly. Thus, it is critical to take precautionary measures to remove unconscious bias.

AI can be taught to filter irrelevant considerations out of the decision-making process; pick the most suitable candidate based on objective criteria.

The algorithms that drive AI don't reveal pure, objective truth just because they're mathematical. People have to teach AI what they consider objective, provide relevant information and indicate outcomes they consider best—ethically, legally, and financially. Continuous filtering and fair and productive system in place can help AI create a diverse workforce.

Let us look at some of the advantages that AI will bring, according to Gartner.

Advantages of Artificial Intelligence over Machine Intelligence

1.

AI will increase accessibility

By the year 2023, the number of employees with handicaps will triple, because of AI and other emerging technologies which will reduce barriers of access. In the USA, just 30 per cent of the labor force members have disabilities and are employed. The rest 70 per cent denotes a gigantic unexploited talent pool, predominantly given that hiring managers are issuing warning signals about talent and skill shortages and the effects on the future of organizations.

Therefore, changes will have to be made in the structure of organizations which may range from the technical (modifying legacy structures to make them more accessible) to the cultural. And there is a lot to be gained as far as providing employment to people with disabilities is concerned. Let us look at some of the figures. Groups that employ people with disabilities have about 89 per cent higher retention rates, couple with a 72 per cent increase in employee productivity, and a 29 per cent increase in profitability. Besides, increased diversity translates into fresh perspectives. Workers with challenges could view product development from an entirely different perspective. This increases the potential for a product which would appeal to a wide client base.

2.

AI emotions drive ads

By the year 2024, AI identification of emotions will impact more than half of all online ads. As acceptance of biometric-tracking sensors continues to rise and artificial emotional intelligence grows, industries will be proficient in sensing customer emotions and use the knowledge to upsurge sales.

Along with behavior and environmental indicators, biometrics enable a profounder level of hyper-personalization. However, brands will need to be transparent about how they are gathering and utilizing consumers' data.

3. **Inescapable AI:** In digitally-savvy enterprises, everything would be embedded with AI. Given the evolving industry, AI-enabled enterprises would be able to sense as well as respond to changes. Enterprises need to hasten better decision-making with AI-powered data analytics, which can cull out insights from any relevant unstructured data, irrespective of origin as well as format and enable unified and context sensitive search in addition to knowledge discovery across video, image, audio, and text data. This will be the engine that will spur data transformation across the industries of the world.

CHAPTER TEN

AI Startups Generating Path-Breaking Innovations

AI Startups Generating Path-Breaking Innovations

New technologies such as Artificial Intelligence (AI) are gaining great traction, particularly in developed countries such as the USA and Germany. With rapid increase in number crunching and data-gathering capabilities of computers, AI applications and tools can only expand in the future. Internationally, AI-driven systems are paving the way for a more collaborative, explorative, and focused understanding as well as leveraging of business data.

Cognitive Code: Leading the way in conversational experiences

Termed as a 'Gartner Cool Vendor in AI for Conversational Platforms,' the SILVIA Artificial Intelligence Suite by the company is considered to be the most advanced, portable, and flexible system in the world for creating immersive conversational experiences. Designed from scratch and through years of innovative approach, this natural language solution is unlike any other. SILVIA is lightweight as well as easy to use. It can also operate in spaces, where other systems may not function.

> *"'The SILVIA platform was designed to be far more than a personal assistant. SILVIA doesn't just converse, she holds a conversation. She can inform, instruct, and learn in a very natural way. And the elegant architecture of the platform puts her capabilities directly in the hands of content developers,' says Chief Product Officer, Alex Mayberry.*
>
> *"With our tools and technologies, developers and content creators can now rapidly build compelling conversational experiences for just about any application, from small indie games to large-scale enterprise business solutions and anything in between. SILVIA has minimal processing and memory requirements, operates across a variety of devices and operating systems, and does not require an internet connection. This means that SILVIA interactions are not only a great experience for users, but are also private and secure," Adds CEO and Founder Leslie Spring"*

Lorose

Brighterion: Exploring New Frontiers in Fraud Management

The innovation-based organization offers one of the world's deepest as well as broadest portfolios for AI and Machine Learning tools to provide real-time intelligence from data sources no matter what the type, volume, or complexity. This is particularly applicable for identity and fraud management.

""Brighterion has deep academic roots; this technology applies to 10 different AI and machine learning technologies. Notable is the use of its Smart-Agents technology, which continually adjusts understanding of behavior patterns of specific profiles of entities without need for rules or model updates. Combined with Smart-Agents, the approach offers continually evolving models with significantly lower operational and modelling costs. It has demonstrated dramatically lower false-positive rates and detection rates in head-to-head comparisons with legacy financial crime scoring models," says Gartner."

In fact, the company's focus on behavioral analysis provides for cross-channel behavior anomaly prediction and detection. The AI solution also identifies low-risk behavior as well as improves the overall model performance. Brighterion's AI as well as Machine Learning-based tools have been applied in cyber/homeland security, real-time cross-channel fraud prevention, onboarding and risk monitoring, biotech, data breach detection, and financial markets.

Cortical.io: AI Innovations in Language Processing

Founded in 2011, Cortical.io, is an innovator in Natural Language Understanding (NLU) and offers solutions which are based on the theory of Semantic Folding, which opens a radically new perspective on the handling of big text data. The innovative organization's software Retina Engine converts language into semantic fingerprints, numerical representations that capture meaning explicitly and carries out computational operations on that meaning. The uniqueness of its algorithm enables one to solve open NLU challenges, such as meaning filtering of terabytes of unstructured text data, real time topic detection in social media as well as semantic searches through millions of documents across languages.

""Since our foundation in 2011, we have been following a neuroscience-rooted approach to the processing of big text data that profoundly differs from mainstream methods... It highlights the disruptive potential of our technology in a market puzzled as to how huge volumes of unstructured text data can be efficiently processed. The uniqueness of our model is confirmed by customers from areas

as diverse as financial services, information technology and the automotive industry," says Francisco Webber, co-founder and General Manager of Cortical.io. "

~~~~~~~~~~$$$$$$$$$$~~~~~~~~~~

CHAPTER ELEVEN

How Artificial Intelligence Is Transforming Industries

How Artificial Intelligence Is Transforming Industries

Artificial Intelligence and Data Analytics are affecting all industries. In fact, there is not one industry that is not affected. Let us look at a few examples: A restaurant In Japan, a restaurant is experimenting with AI robotics technology which will allow employees with mobility challenges to remotely pilot robotic waiters. Companies like Microsoft, Ford, and JPMorgan Chase are hosting virtual career fairs which are tailored to the requirements of diverse candidates. Enterprise Rent-A-Car has been able to implement braille-reader technology in the booking system for the help of blind employees.

Currently, automotive as well as mining industries use wearables to increase employee safety, on the other hand, travel and healthcare domains use technology to maximize productivity. As technologies continue to evolve, organizations can begin to understand how physical augmentations could be used for personal and professional lives.

Artificial Intelligence: Technology for the Future

Research and analytics firm Markets and Markets' report "Artificial Intelligence Market by Technology" predicts that the AI market would cross USD 16.06 billion by 2022, growing at a healthy CAGR of 62.9% from 2016 to 2022. Additionally and surprisingly, most of the cutting-edge AI technology is not only related to robots, but also to contemporary technologies to power business processes and operations and resolve industry disputes.

A recent report by McKinsey Global Institute points to the fact that companies which are new to the space can learn a great deal from the early adopters, who have invested millions and billions in developing AI and are now beginning to reap a range of benefits. For instance, retailers across the globe are relying on AI-powered robots in order to run their warehouses.

Similarly, those in the energy and utilities space are using AI to forecast electricity demand. Automakers are harnessing the technology in self-driving cars. The technology is witnessing a gamut of use cases as well as interventions. This shifts technology's position from being only an enabler to a disruptor of traditional business models. As per a PwC FinTech Trends Report (India), global investment in AI applications reached $5.1 billion, up from $4.0 billion.

In fact, these new technologies are being viewed as cornerstones of a Digital World. AI tools can provide the flexibility and scalability needed for enabling digital transformation for enterprises in the banking, finance, and fintech space. For example, the banking and financial sectors are leveraging the digital to improve customer experience as well as bottom lines. However, transformational strategies need to be molded around technology benefits in order to leverage maximum profitability.

How Artificial Intelligence Is Changing the Face of Banking

Artificial Intelligence Is Changing the Face of Banking

If one thinks of the banking sector, there has been an increasing adoption of digitally disruptive technologies as compared to the other sectors. This is despite the fact that traditionally, banking has been a manpower-intensive sector, with operations that require human involvement. The banking sector has been an early adopter. Digital transformation is affecting banking in terms of products, processes and strategies, including the use of predictive and cognitive capabilities.

Banking is a data intensive business and AI provides predictions and recommendations through processing huge volumes of multiple data sources. Banks know their customers deeply and their data is valuable as they have access to financial details of customers.

AI can make recommendations about products and services to customers by processing such details. Also, AI can automate repetitive tasks, saving on time, effort and cost. AI is finding wide ranging use cases in Capital Markets, Consumer Banking and Insurance. Robo Advisors, Risk Management, High Frequency Trading, Cyber Security, Intelligent

Predictions, Fraud Detection, and Recommendations.

Technology is steadily seeping into the operations and thus efficiently cutting down redundant tasks. This makes operations hassle-free and impactful as well as create a leaner system to work on. Cloud computing, mobile-first, and digital dashboards have already started becoming the norm.

> ""*Currently, processes that were earlier getting done manually are now being done through AI/ML tools, including processes such as robotic process automation (RPA). Intelligent automation is where the greatest activity is and this trend should continue. Data Analytics can be optimally leveraged if the input and algorithms are in place. Interfaces can be enabled or even made voice-based. The technology is already there, the challenge is that the practitioners don't know enough about it. Moreover, technology takes time to deploy. In my estimation, within 3 years, a number of these technologies should get deployed and come into use," says another, who has held top positions in several banking organizations.* "*

At the core of the banking business model is to make every customer interaction digitally enabled, whether it is account opening via a tab or other functions. Machine Learning and Deep Learning are already impacting the domain of Risk and Credit Assessment and insurance. Globally, banks have begun providing smart wallets to customers.

AI enabled smart wallets look at customer's spending habits and learns from their behaviour to provide smart advise and recommendations for future spending. It boosts savings and responsible spending through predictive alerts and recommendations. AI can identify if a customer is likely to switch products or services. This can help banks offer more suitable products, leading to customer retention.

> ""*In the Risk Assessment, Credit assessment and Regulatory areas, if you apply for a loan, in a conventional way then a home loan or a personal loan takes couple of weeks or may be even more to clear all kinds of credit checks before approving such loans. With AI, the processing time will come down due to faster credit assessment.*
>
> *We are experiencing change in the interface of banks with customers. They are increasingly shifting to Chatbots, Robots and*

Humanoids as the first line of interface with customers in order to enhance customer experience. Facial analysis to detect ATM frauds. AI advisers called Robo-Advisers are moving towards providing customers with round the clock intelligence," says an expert.
"

Due to increased digital footprint, financial crimes are growing exponentially. Orthodox approaches are ineffective as digitization is increasing and hackers and fraudsters are getting more sophisticated and innovative. Machine learning can help programs learn from past events and help distinguish between the usual and unusual transactions. An amalgamation can lead to the creation of a wide-ranging solution. AI can help in automating financial crime detection in a timely, efficiently and cost-effective way.

Examples of a few AI and ML solutions: Fraud detection in high volume transations; Money Laundering (identify source and destination of the money entering legitimate financial systems); and Credit Card fraud detection. Smart text and speech analytics solutions such as customer email analytics using NLP (Natural Language Processing) can identify emails such as escalation emails, follow-up emails or emails seeking information and presenting them in the form of a dashboard with clearly identified categories. Such tools are fast getting adopted.

Speech analytics and processing solutions using NLP & deep learning can be leveraged for monitoring and improving customer care interaction quality, real-time conversational assistance and guidance to agents during a live interaction with customers. Text processing and Natural Language Generation (NLG) can automate statement and report generation like financial reports and credit card statements and other such routine activities.

Unstructured document processing can be automated through AI tools as banking is a document-intensive industry. This can be helpful in loan processing or insurance claim settlements, and reduce costs and turnaround time and boost efficiency for processing. Customers can be provided services through a number of digital channels such as social media and others in the form of ChatBots and Virtual Assistants as customers want to avail the same banking experience over multiple digital channels of their choice and ease.

It can also help in customer segmentation and recommendation for the right product at the right time with the aid of efficient and targeted marketing campaigns, customer lifecycle analytics as well as predictions which can help in improving and sustaining relationships with customers. This can improve bottomlines and improve profitability helping banks strategize retention plans accordingly and acquire competitive edge. In the sphere of smart KYC as well as building customer 360-degree profile, intelligent cognitive tools and solutions can make a mark. AI-based authentication systems such as multi-factor voice identification, face recognition and liveliness detection can be used for authentication and financial transactional activities.

Although there are regulatory challenges for such implementation in some of the countries it's going to be the de-facto in the near future. AI is already transforming the payments landscape. In the future, intelligent payment systems can be used within the customer journey. This can be transformative shift in paradigm to non-interventional way of payment and fulfilment of transactions. AI technology can make trans-border transactions smooth.

Intelligent identity management as well as smart security monitoring solutions are becoming popular. Tools based on computer vision, speech-based techniques, cameras and sensors can beef up security and detect unauthorized access, criminal activities inside bank ATMs, branches or lockers. The cognitive-powered systems can generate real-time alerts as well as notifications to prevent mishaps. Facial stress analysis can be used for detecting ATM frauds.

AI tools and chatbots are already being deployed by several leading players to enhance customer experience. There is no paper involved and no filling of forms. The benefits are myriad. The customer spends less time along with the fact that there is no rework at the back end. Plus, there is no data leakage or loss. There are AI/ML tools these days that will crunch all the data backend and provide you with insights into customer acquisition, the most capital intensive part in the financial world.

No wonder, banks are already leveraging such tools in the area of customer targeting and how to design marketing strategies. Basic clustering is manual in nature currently. It basically classifies products category wise. For example, it will slot products for Category A, B, or C. However, all that is changing now.

For example, if you placed a request for a check book at a branch, it would capture the data. Even if you went on Internet banking the next moment, it would track the request, since they are all interconnected.

There is visibility and the customer feels in command of his transactions.

> ""*The biggest use case of Artificial Intelligence and Machine Learning, which everybody is focusing on right now, is in the underwriting process for loans. This is where the action is right now. The underwriting process is very chaotic currently. Humanly, it is not possible to do underwriting for loans for say Rs. 50,000. This is one area that organizations are working on to create the best value out of these technologies. However, several banks and fintechs are already at several stages of designing the processes. The second use case is fraud management. Manually, it is not possible to figure out trends. Even if one is able to figure out, how does one stop it from happening? This is where technology can be of great use," says Sameer Singh Jaini, CEO and Founder of The Digital Fifth.*"

With the coming of AI/ML tools, there is a lot of apprehension over jobs losses as these technologies go about digitally disrupting industries and sectors.

> "*In larger entities, we should expect 30-50% of headcount decrease. There are a number of processes that can be automated. Some of the typical mundane roles will go. The banks will shrink in size in terms of the number of people. Most transactions will be instantaneous, says a banking expert.*"

Machine learning tools are providing insights into how to approach a customer or how to elicit a favorable response by tweaking the existing processes. The best part is that this wave is favorable for startups and those wanting to start out afresh as they are not burdened with legacy systems and infrastructure. There are a lot of use cases of such technologies that the industry can take advantage of.

AI has also been finding use in several personal finance applications, such as Mint or Turbo Tax. These developments are disrupting financial and commercial institutions. Such applications gather data and provide advice.

Programs like IBM Watson have been applied to the home buying as well. These days, it is software which performs most of the trading on Wall Street.

How AI is Impacting the Insurance Sector

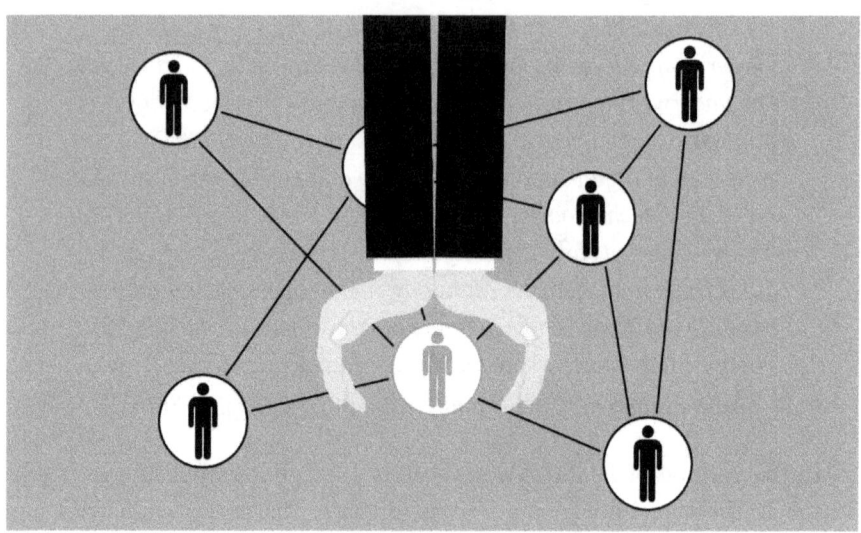

AI in the Insurance Sector

Sophisticated AI and Analytics tools together with Cloud-based enterprise software can result into a focused, collaborative, and explorative systems. Business processes, which always depended on human agency, involving repetitive tasks such as checking invoices, perusing résumés, travel expenses, etc. can exploit them to their advantage. Insurance is one of the sectors that is leveraging AI technologies to the hilt.

Self-learning AI algorithms are helping insurance companies detect patterns and solutions in data thus attaining new levels of efficiency.

At an event, Tarun Chugh, MD & CEO at Bajaj Allianz Life Insurance spoke about how Bajaj Allianz Insurance utilized this tech. Giving examples, he talked about a newly developed chat bot which was proving to be an astounding success. The AI-enabled self-learning tool registered 4 lakh queries and completed 45,000 transactions, all within only 45 days.

" "End to end business processes are being handled by our newly developed AI bot. What once took three days for the most efficient of processes, such as issuance of policies now takes three hours. In fact, the number has come down to three minutes and we are planning to get it down further to real time, said Tarun Chugh, MD & CEO at Bajaj Allianz Life Insurance. "

Lorose

AI in Healthcare

Artificial intelligence makes its way to MacDill dermatology.

The biggest applications are reducing treatment costs and improving patient outcomes. Enterprises are applying machine learning to make better and faster diagnoses than homo sapiens. One of the best known healthcare applications and technologies is IBM Watson. It comprehends human language and can respond to questions. Watson mines patient data to form a hypothesis, which is then presented along with a confidence scoring system. Other applications include chatbots to help schedule appointments or help patients during billing or virtual health assistants which provide basic health feedback.

AI in Business and Education

ML algorithms are getting integrated into analytics as well as CRM platforms to discover information on how to serve customers better. In fact, chatbots have also been built into websites to provide real-time service to customers.

AI tools can help automate grading and giving educationalists more time. The AI systems can also assess students and adapt to their needs, thus helping them work at the pace of their choice. AI tutors are capable of providing support to students and thereby ensuring they are on track. In fact, AI can change how and where pupils learn, perhaps even replacing some teachers.

AI in Law and Judicial Systems

Artificial intelligence and law is a subfield primarily connected with resolving legal informatics problem together with original research into those problems. It also contributes to tools and techniques in the context of legal problems.

The discovery process which includes tedious sorting and sifting of documents can be quite a task. Automating such time-consuming process can lead to a more efficient judicial system. There are several startups which have started making question-and-answer computer assistants to help in the process.

For instance, in theories of legal decision-making, particularly models of argumentation, knowledge representation and reasoning; models of social organization and multi-agent systems; case-based reasoning; and storage and retrieval of huge amounts of text-based data. This has led to the formation of information retrieval systems and intelligent databases.

Such tools are often used in dispute resolution platforms that utilize optimization algorithms and can be leveraged for modeling legal ontology. There are other applications as well.

AI in Manufacturing Industries: The Rise of Smart Factories

AI adoption in manufacturing processes

"Co-founder of Google Brain and Coursera Andrew Ng says: AI would carry out manufacturing, shorten design time, quality control, and reduce materials waste, improve production reuse, perform predictive maintenance, and more."

The manufacturing sector has always been at the vanguard of integrating robotic systems into their workflow. Now, AI is transforming manufacturing in ways unconceivable. Industrial robots have been used to perform tasks on the factory floor. According to a Global Market Insights report, the market size for Artificial Intelligence (AI) technologies in the manufacturing sector has surpassed $1 billion and is estimated to grow at a CAGR of over 40 per cent from 2019 to 2025.

However, some experts say that the sector is yet to fully embrace the tech and insist that it is moving slower than expected as compared to other sectors and considering its market and size.

~~~~~~~~~~$$$$$$$$$$~~~~~~~~~~

CHAPTER TWELVE

Major forms of AI Tools Being Used in Industries

There are various different kinds and forms of Artificial Intelligence and Machine Learning tools that are being used across industries.

Chatbots Enhancing Productivity

Chatbots Enhancing Productivity

Chatbots are currently the most popular face of AI and influence all areas where there is communication between humans and machines. For example, car maker KIA talks to 115,000 users per week, or Lidl's Winebot Margot provides help on which wine to buy in addition to guidelines on food pairings.

Chatbots could be text- or voice-based. Or they can be a combination of both, relying on scripted responses involving few numbers of people. There are several common applications which exist in HR, IT help desk as well as self-service. It is customer service where they are having the greatest impact, remarkably altering the way customer service is managed.

Automation and Robotic Process Automation (RPA)

Automation and Robotic Process Automation (RPA)

Automation is a technology through which a process or a procedure can be carried out with minimal human assistance. This includes automation of various control systems for operating equipment like processes in factories, machinery, boilers and heat treating ovens, telephone networks, steering of ships and aircraft, etc.

Blue Prism is credited for coining the term RPA. It uses AI and ML capabilities to handle repetitive and high-volume tasks which previously required humans to perform such as queries, calculations and maintenance of records and transactions. It is a digital workforce that follows rule-based business processes and interacts with systems in the same way as existing users do.

The RPA market is expected to reach $3.11 billion by 2025. It has been used to handle the General Data Protection Regulation by Tokio Marine Kiln, by nutrition company Fonterra to fix quantity mismatches and by Milaha for entry, processing and transfer of data. The main companies in the RPA market include UIPath and Automation Anywhere.

Machine Learning and Deep Learning

The science of getting a computer to act without programming. Deep learning is a subset of machine learning that, in very simple terms, can be thought of as the automation of predictive analytics. Machine learning has the potential to crack business problems like dynamic pricing, personalized customer treatment, supply chain recommendations and suggestions, and medical diagnostics. ML utilizes mathematical models and algorithms to source knowledge and patterns from the data.

Such an adoption of ML is increasing across organizations to make sense of the exponential explosion in data volumes and advancements in the compute infrastructure. Today, ML is being put to use in numerous fields and industries for finding new solutions.

American Express uses analytics as well as algorithms to detect fraud in real time and save millions in losses. Volvo leverages data to predict when specific parts might start failing or when they would need servicing, thus enhancing vehicle safety.

Machine Vision and Computer Vision

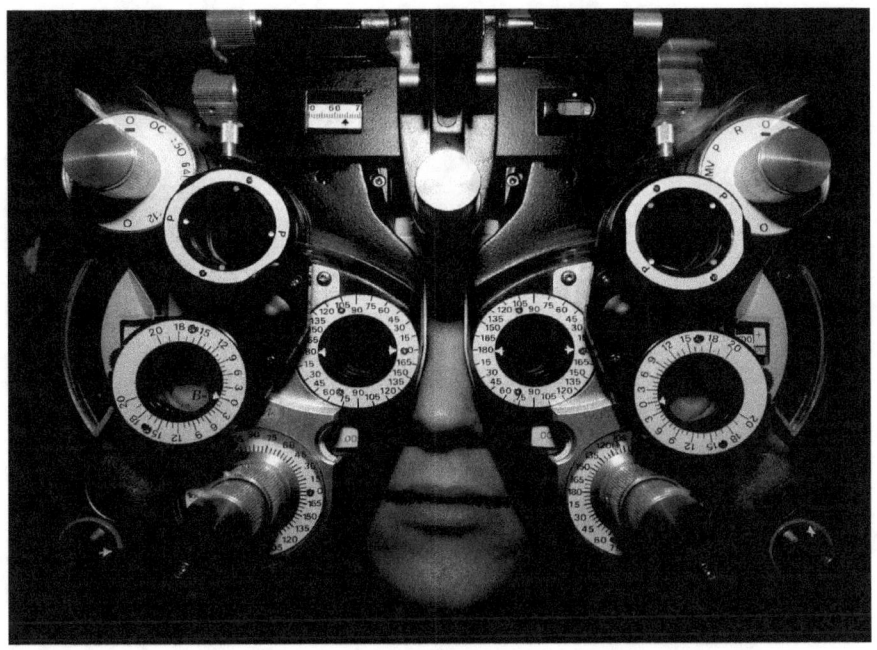

Machine Vision System

This technology allows computers to see. A computing device is able to inspect, evaluate and identify still or moving images. Machine vision systems (MVS) and methods are used to provide image-based automatic inspection and analysis for applications such as robotic guidance, process control, and automatic inspection, and is usually used in industry.

It captures and analyses visual information through the use of a camera, analog-to-digital conversion as well as digital signal processing. It is quite similar to surveillance cameras, but also provides automatic image capturing, evaluation and processing capabilities. Some experts consider it to be similar to human eyesight. However, machine vision can be programmed to see through walls.

The tech is used for applications ranging from medical image analysis to signature identification.

Natural Language Processing (NLP)

Natural Language Processing (NLP)

This is the processing of human language by a computer program. Natural language processing (NLP) is a subfield of linguistics, computer science, information engineering, and artificial intelligence concerned with the interactions between computers and human (natural) languages. The ultimate objective is to read, decipher, understand, and make sense of the human languages in a manner that is valuable.

Such programs can process large amounts of natural language data including speech and text with the help of software. This has been in existence since many years and basically grew out of the field of linguistics along with the rise of computers and computing power.

This sub-field of AI can also process unstructured text and extract data from it. For example, email spam detectors (which are some of the oldest examples of NLP) check the subject line as well as the text of an email to decide if it is junk. Current approaches to NLP are based on machine learning. NLP tasks have come to include speech recognition, text translation, and sentiment analysis.

The Age of Industrial Robotics

The Age of Industrial Robotics

Robotics is   an inter-disciplinary branch   of engineering that includes computer science, mechanical and electronic engineering, among others. It deals with the design and use of robots for control, information processing, and sensory feedback.

These technologies replace humans and imitate their actions. Robots can be utilized in several situations. Today, they are confined to use in dangerous environments (such as bomb deactivation), hazardous manufacturing processes in which humans cannot survive (for example, in high heat, in space, under water, or cleaning up of hazardous materials or radiation). They can be used to execute jobs difficult to accomplish for human beings.

They often find use in car production assembly lines or by NASA to move large objects in space. Research is going on to use machine learning to build robots which can relate to social settings.

They can ideally take on any form but often resemble humans in appearance to help in their acceptability. Such machines try to replicate lifting, walking, speech, and other human activities. Several robots are inspired by nature also, leading to the domain of bio-inspired robotics.

Intelligent applications Will be a Game Changer

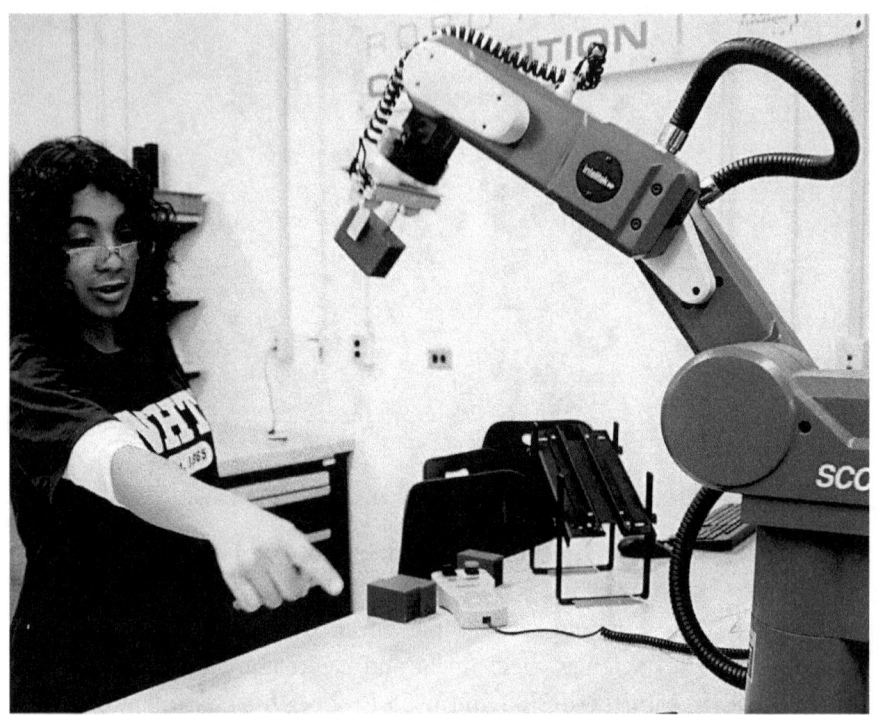

Intelligent applications Will be a Game Changer

Corporations' inclination for procuring AI capabilities is now transitioning in favor of acquiring them in enterprise applications. Intelligent applications are enterprise applications which will come with integrated or embedded AI technologies in order to support or replace human-based activities by means of data-driven insights, intelligent automation, and recommendations to increase productivity in addition to decision-making.

These days, enterprise application providers are already incorporating AI capabilities and technologies in their offerings together with creating AI platform capabilities. Such capabilities have found space in domains ranging from customer relationship management to enterprise resource planning to human capital management and workforce productivity applications. Chief

Information Officers (CIOs) need to challenge their software providers to produce a blueprint of how they are embedding AI in order to add business value. This could include elements such as intelligent processes, advanced analytics, and advanced user experiences.

Augmented Intelligence

Augmented intelligence is primarily a human-oriented partnership model with people and AI collaborating in order to improve cognitive performance. This incorporates AI's role in advancing human competencies. Such a collaboration between AI and humans refines existing knowledge about errors in repetitive work.

The new tech also has the potential to improve customer interactions, citizen services as well as patient care. The objective is to improve efficiency with automation, on the other hand complementing it with common sense to manage the risks of decision automation.

~~~~~~~~~~$$$$$$$$$$~~~~~~~~~~

CHAPTER THIRTEEN

AI Tools Can Enhance Customer Experience & ERP

AI Tools can Enhance Customer Experience

According to the research company Markets and Markets report "Artificial Intelligence Market by Technology," it is expected that the AI market would be worth USD 16.06 billion by 2022, growing at a stupendous CAGR of 62.9% from 2016 to 2022. It is affecting all industries. Similarly, its impact on the world of enterprise technology is being felt in myriad ways.

"The more we digitalize, the more our prosperity is secured," says Wolfgang Wahlster, who leads the German Research Center for Artificial Intelligence (DFKI).

A scientist with Google Research, Chris Olah, who is also the author of an AI research paper called "Concrete Problems in AI Safety," says, "The authors believe AI technologies are likely to be overwhelmingly beneficial for humanity but we also believe it is worth giving serious thought to potential challenges and risks," adding: 'We strongly support work on privacy, security, fairness, economics and policy." "

Lorose
What is ERP software?

The Enterprise resource planning (ERP) software is the backbone of the modern enterprise. This software allows a company to use a system of integrated applications and tools to help them manage the business and automate back office functions such as services and human resources. And it is getting majorly impacted by the evolving AI technology.

AI tools revolutionizing ERP software systems

AI is revolutionizing the ERP software domain and therefore radically transforming the way businesses function. Together with the data mining capability of advanced self-learning enabled AI systems, automation of ERP software systems could bring great value addition to a business.

The AI-enabled ERP software can influence the core of an organization's day-to-day business function and processes by cutting down on operating expenses, streamlining tasks that are routine and removing human errors. AI tools can be incorporated into ERP systems so that they can learn processes needed for improved competencies to enhance the effectiveness of overall business processes. Advanced ERP technology is helping organizations discover colossal amounts of structured data. With every passing day, they are getting smarter at learning new patterns and quicker at converting the data deluge into insights.

Workforce expertise and skill can be optimized through AI tools, increasing the overall effectiveness of ERP as the focus shifts in favor of non-routine, analytical, logical, and creative tasks. There is another positive aspect to it as well: AI can liberate workers from manual and monotonous interactions with the ERP software by triggering an evolution of self-driving and semi-intelligent business solutions. AI tools can also be used for ERP software maintenance.

AI-enabled ERPs employ a digital assistant (DA) in order to help technicians perform the root cause analysis for maintenance issues. The provisions can record information and details of performance, maintenance history of a device, and technical structure. The DA tool can extract as well as source information from core ERP systems.

AI tools can mimic functionalities of a human mind, such as behavioral changes as well as learning and can apply emulated tactics to unstructured data in the ERP. Though there may be apprehension in the adoption of AI tools, there are varied and extensive benefits that can be drawn out of it. It will eventually help organizations improve the primary focus of their businesses, which is to serve customers well.

AI Tools Can Provide the Edge to Application Monitoring

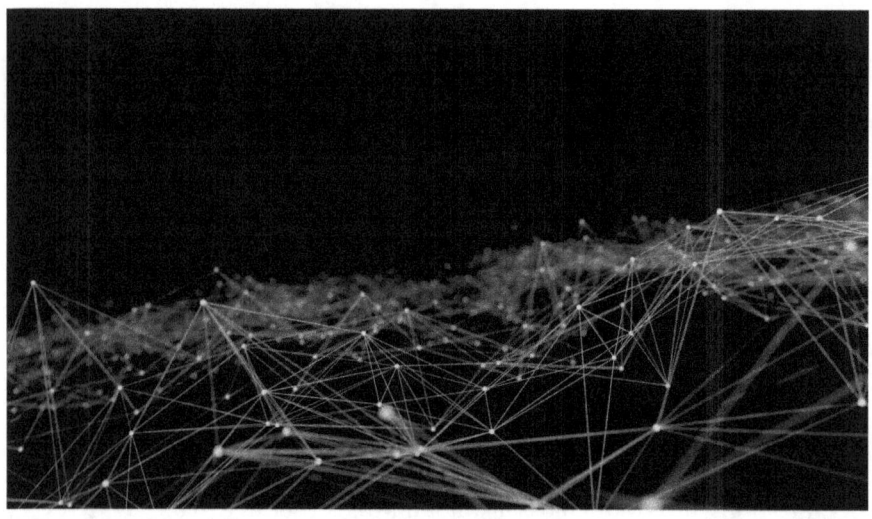

AI Tools Can Provide the Edge to Application Monitoring

There is an explosion in the number of applications. In fact, we are already in what many experts call an app economy. In such a scenario, the concept of "DevOps" enters the picture. The name comes from streamlining software development, which is basically the act of writing computer code, with IT operations. These teams test, deploy, as well as monitor new applications or platforms. This is now an important part of the operations of many companies these days. Companies can no more afford to be driven by processes that can be an obstacle to business growth.

The biggest benefit of Devops is the ability to make changes as we go along. It allows businesses to be flexible when it comes to requirements. Devops offers you to work together in an agile manner. Through this collaboration between development and operations, the effort in meeting and delivering business requirements as well as the time to market gets reduced. Another big advantage is that you don't have to rewrite code. You get microservices available through the community.

AI tools can tremendously help in application monitoring. Besides, they can help organizations analyze the terabytes of data coming out of the applications and make sense of it in real time. Artificial Intelligence, in particular, can help in automating such data heavy lifting and proactively identifying problems and pinpointing the underlying root cause.

Moreover, a highly converged IT infrastructure and the distributed application environments are quite a challenge to manage as data moves at a fast rate. The sheer velocity, volume, as the diversity of information cannot be controlled by just using traditional tools and methods like watching dashboards, manually analyzing data sets, or responding to alerts. Continuous innovation demands that the next-gen and advanced solutions are equipped with modern tools.

In fact, AI is not an option anymore but a necessity in complex application environments and deployments. The analysis that may take experts hours or may be days, can be done in milliseconds. No wonder, even tech behemoths like Microsoft and Apple realize the power which AI can bring to their platforms. An AI engine can add value to the application stack, particularly in the troubleshooting of an application, with the true root cause analysis (RCA).

" "Real-time analytics and AI automatically detect performance and availability issues and pinpoint the root cause, so that application problems can be resolved before the customers are impacted. There is a view of the entire digital ecosystem in high fidelity: every transaction from every user is delivered by our full stack, single agent. We incorporate third-party data, including time series, unstructured data, events, relational data, and more," says Steve Jobson, VP, Dynatrace LLC.

He says that real-time analytics and predictive analytics are keys to the success of organizations. The company is a USA-based AI-powered, full stack solution provider for hybrid multi-cloud enterprise. The company is in Gartner Magic Quadrant for eight years and has developed an AI Analytics engine which captures every container, every interaction, and every piece of code to predict and self-heal the digital ecosystem."

In all this, the role of data cannot be emphasized enough. The AI engine can only be as good as the data which feeds it. It provides application monitoring, user experience management, and network and application monitoring solutions, such as Web performance monitoring and mobile application monitoring, along with comprehensive data center monitoring.

The AI engine automatically detects virtual and physical relationships and changes. This allows for correlation and it gives you true causation. AI engine uses different algorithms and statistical models such as the Graph Ranking Methodology, Multidimensional Baselining, Temporal Correlation, Holt-Winters Exponential Smoothing, etc. In today's distributed application environments, an AI engine can show relationships in real time and can provide the context through which one can tell not only what happened but why it happened.

Artificial Intelligence can Enhance Digital Customer Experience

Artificial Intelligence can Enhance Digital Customer Experience

Boxever, a company, makes products that heavily rely on AI and ML to enhance the customer experience in the travel industry and conveys micro-moments or experiences to please customers. This significantly improves customer engagement, helping them find new ways and make memorable journeys.

"What lies at the core? It's clearly the customer experience. Today the customers expect everything to happen seamlessly; in real time. Customers expect that as a bank or as the brand, you must know and understand them in and out. To do that, organisations need to leverage the data they collect. Customers leave a lot of information. It depends on organisations how they mine the data to make informed decisions and ascertain customer's choices and make their lives simpler. These paradigmatic changes are now becoming the core of business that is forward looking and aiming growth. Unless you keep customer experience at the core of business, you will have a challenge staying relevant. To do or not to do is not a choice anymore. If you have to stay relevant, you have no choice but to take

this path, says Deepak Sharma, a leading CDO and tech leader."

This is true. Customers today are not as tolerant and forgiving as they were a few years ago. While customer acquisition cost has climbed down, customer retention and engagement are bigger challenges. Unlike past, customers have many options today. They can easily exit any network, system or platform if their demands are not fulfilled instantaneously. The modern customer expects the best-in-class experience. Whether it is about shopping, entertainment or dining, experience is all that counts.

"*About 80 per cent of consumers will move from a product to another if they are not satisfied with their experience within three months, says an expert.*"

That is the real problem. People are forced to spend huge amounts of money in customer acquisition and they could lose it. Companies need to look at software functionalities and differentiation. Customers demand it to work perfectly. They don't expect failures at all. There was a time when a down time with technology was tolerated in good humor but not anymore.

"*"Moreover, it is not the application any more. It is much more than that. It is the devices, the connectivity, the whole experience. And everything is software driven. Lot of companies are under pressure to perform at this pace and with high quality," says another expert.*"

~~~~~~~~~~$$$$$$$$$$~~~~~~~~~~

CHAPTER FOURTEEN

Bolstering Cyber Security, the AI Way

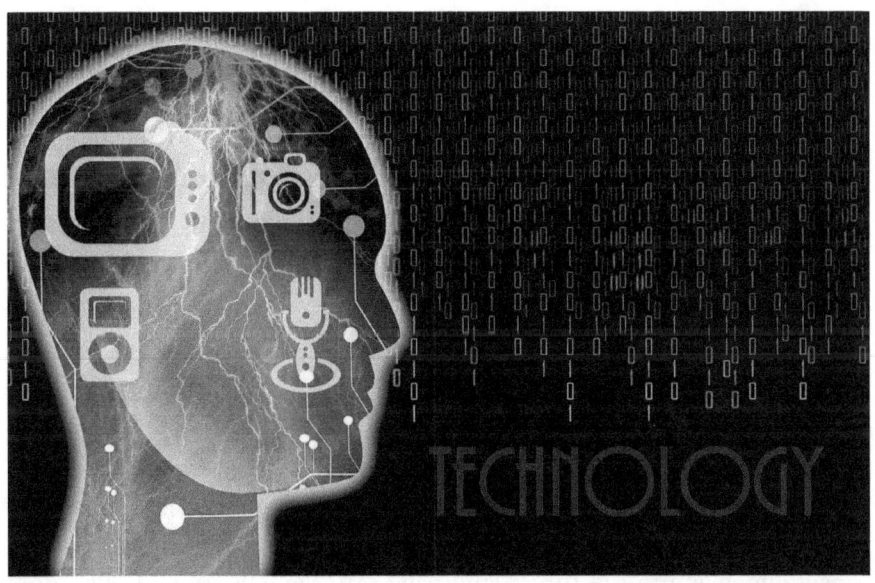

AI for Cyber Security

Digitalization brings a set of unique challenges. The new-age digital platforms are not free from vulnerabilities, in fact, they sometimes end up increasing them. We live in an App world today. The threat quotient increases as we move towards an ever-increasing use of applications. The attack surface has significantly increased, giving greater leeway to cyber criminals.

Cyber threats continue to grow in frequency and sophistication as the threat landscape quickly evolves. In fact, modern day attacks are not only capable of crippling the operations of an organization, but can cause irreparable damage to a company's reputation as well as costing leading to losses running in billions of dollars.

There are examples a plenty of the increasing severity of such attacks. Global pharma giant Merc lost $300 million dollars in one quarter only due to a cyber-attack. Organizations have to take a holistic, analytics-driven approach to secure identities, applications, and data. The security ecosystem requires data, tools, people, and processes. The information flow from the SOC to the boardroom has to be seamless for better risk management. Security teams need to evolve rapidly to stay ahead of the attackers.

In the constantly changing threat landscape, digital transformation technologies such as Artificial Intelligence are being increasingly looked at to mitigate and stave off the threat factor. There are a lot of use cases of AI which can help security professionals take the right decisions. Companies and their chief information security officers (CISOs) can confidently adapt to the fast changing threat landscape, with the help of the latest technologies available such as AI.

> *""This is a cat and mouse game where we are trying to update our security devices and at the same time some new threats are coming in. So, with the influx of new technologies such as AI and ML, threats could be automated," says Anshuman Pund, Head - Information Security & ISO Compliance, IDBI Intech Limited."*

In fact, the explosion in data due to digital transformation has also added a new dimension to the threat landscape. Attackers continue to get more sophisticated. Attackers spend significantly less time and resources to bypass security checks and tools. Security teams need to evolve rapidly to stay ahead of the attackers.

> *""With the increase in zero-day attacks that constantly morph and evolve, it is becoming extremely challenging for the existing security solutions to effectively tackle them. There is a distinct need for solutions that keep pace with the attackers. AI has the capability to properly perceive the cyber threat environment to identify events*

and take appropriate action. It is particularly good at recognizing patterns and anomalies within them, which makes it an excellent tool to detect threats."

Machine learning is often used with AI. It has the ability to "learn" on its own, based on human input and results of actions taken. Together with AI, ML can become a tool to predict outcomes based on past events. With the attacks becoming more and more voluminous and sophisticated, there is a growing need to become proactive and use predictive models to thwart potential threats; which can be facilitated through the use of AI and ML," says another cyber security chief, adding: In fact, AI and ML are slowing becoming integral parts of the cyber security set-ups these days.

For AI to be successful, companies will need to standardize processes to gather and analyze data across all devices / applications. Voice and Visual search-based queries need to grow. Modern day security will have to evolve to take full advantage of AL and ML technologies.

Another cyber security expert adds:

" "Artificial Intelligence now has taken the space where no ordinary analytics can deliver results. There are a lot of use cases in AI which help security professionals take the right decisions. The simplest one is early fraud detection. You could also write an AI on user behavioral analytics to identify anomalies of user behavior, which would then be used to identify potential disgruntled employees and thereby preventing insider threats. So, the scope of AI is unlimited and is in a nascent stage and is still developing," says Amit Pradhan, CTSO, Chief Privacy Officer and SVP - Tech Security, Vodafone.

~~~~~~~~~~$$$$$$$$$$~~~~~~~~~~"

CHAPTER FIFTEEN

Is the Age of Software Revolution Truly On?

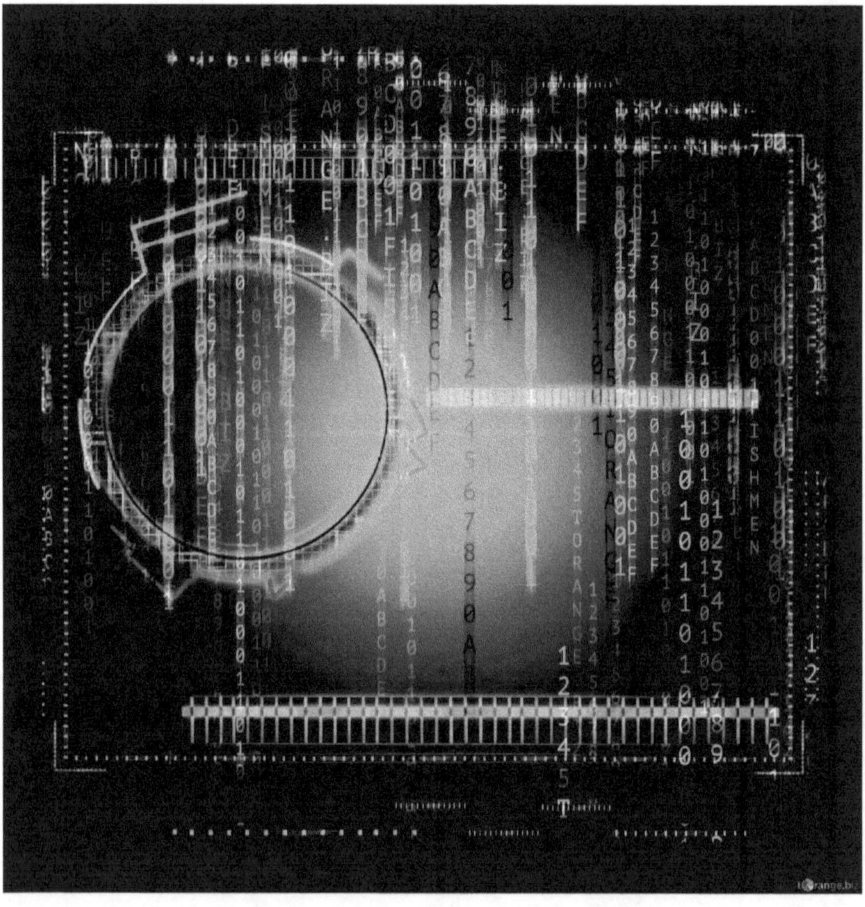

The Age of Software Revolution Dawns

The general inclination towards AI can't be seen in isolation. It is part of the larger agenda of digital transformation. General Electric (GE) made an investment of over $1 billion to build a software "Center of Excellence" in California, USA in order to cope with the exponential data explosion from intelligent machines.

CEO Jeff Immelt declared that GE needed to become a software as well as an analytics company or the organization could risk seeing its own hardware products become commodities as competitors moved in.

> *""On our current trajectory, GE is on track to be a top 10 software company," said former CEO Jeff Immelt. And the astonishing fact is that GE has been a manufacturing company for the past 124 years. Today, companies define themselves in terms of software, he adds."*

This is what is being termed as the software revolution by many industry pundits. No matter what the vertical, all companies are defining themselves in terms of software, both at the customer end as well as in terms of business processes.

> *""We're no longer selling customers just a jet engine, a locomotive, or a wind turbine; we're bringing data and actionable solutions along with the hardware to reduce costs and improve performance." Said Marco Annunziata, Chief Economist at GE, in a media interview."*

On the other hand, a legacy company such as Walmart is struggling to keep up with the changing times and particularly the online market. They are also forced to pull up their socks and get into the race.

Customer Experience Will Decide the Winner of Tomorrow

Of late, they have tried to turn around things for themselves by working around innovation and partnering with software giants like Microsoft and taking their ecommerce platform to the market.

The reason why Walmart and all such companies are getting into the race is because otherwise they risk becoming redundant. As global leaders set the pace, people are upping their expectations.

Digital Transformation Needed to Maintain Business Leadership

They wouldn't settle for anything inferior and for this they need cutting edge modern emerging technologies such as AI, machine learning, blockchains and robotic process automation (RPA) which have significant potential to streamline business operations and enhance customer experience. The potential of such technologies to improve user experience as well as address the demands and expectations of the new-age customer is huge.

Typically, the younger generation expects high speed, agility, and new functionalities as a given. The tolerance level of such customers is very low and they move their feet very quickly.

> *""According to our survey, 80 per cent of consumers will move from one product to another if they were not satisfied with their experience within a window of three months," says Steve Jobson, VP, Dynatrace LLC. "*

According to Jobson, it is not just about the application now; it is much more than that. It is the device, the connectivity, the whole experience, with everything being software driven and that a lot of companies are under pressure to perform at this pace and with high quality. With companies and enterprises joining this race, the software revolution is well and truly on.

~~~~~~~~~~$$$$$$$$$$~~~~~~~~~~

CHAPTER SIXTEEN

Challenges to Adoption of Artificial Intelligence

Challenges to Adoption of Artificial Intelligence

Currently, the roadblocks that the technology faces basically stems from the fact that companies are not able to properly execute the AI in their business processes because the field needs research, innovation, and critical and creative thinking to be fruitful. This will also require significant investments upfront prior to the AI products becoming commercially viable.

> *"Earlier advanced AI functionalities were expensive and exclusive due to undeveloped tools. AI algorithm constructions required highly skilled workforce, generally not available at a number of software companies. CPU-intensive computing and huge amounts of data storage were needed for pattern recognition with respect to the complex algorithms," says an expert.*

Research into AI tools needs to take into consideration several factors including engineering and production specifics including capacity planning, material requirements, infrastructure requirements and costs, as well as product design. In fact, many Indian technology startups have started providing AI solutions, on the back of investors' willingness to see the technology through.

Currently, many organizations are unable to envisage the benefits of AI tools and systems. However, the government can adopt proactive measures to overcome such obvious initial hiccups and boost the adoption of AI in the country. All said and done, SMEs and corporates should approach the technology with an open mind so that they are not 'also-rans' in the race of technology and development.

Security and ethical concerns

While AI tools present a range of new functionality for businesses, the use of artificial intelligence raises ethical questions. This is because deep learning algorithms, which underpin many of the most advanced AI tools, are only as smart as the data they are given in training.

AI is Not Bias Proof

Because a human selects what data should be used for training an AI program, the potential for human bias is inherent and must be monitored closely. AI tools and technology are prone to biases (both implicit as well as explicit), general algorithmic complexities and logic gaps. Such biases can affect whole population groups. This has the potential to severely impact the AI's purpose.

The application of AI in self-driving cars raises security and ethical concerns. Cars can be hacked. Plus, when an autonomous vehicle gets involved in an accident, liability is still unclear. Autonomous vehicles may find themselves in a position where an accident is unavoidable. This can force the program to make an ethical decision on how to minimize damage.

Another major concern could be the potential for abuse of AI tools. Hackers are using sophisticated machine learning tools to gain access to sensitive systems. This is really opening up the threat landscape (more on this in the chapter on Cyber Security.). Deep learning-based video and audio generation tools can be used to create so-called deepfakes, which are primarily fabricated videos of public figures saying or doing things that never took place.

Regulations for AI?

""*What is the algorithm for morality? Would it not change where cultural differences are far right or left? What is considered "moral" in India is not the same as that in Ohio. Saudi Arabia much different than Massachusetts. So the writers of the software end up acting like the parents of the computer program. In a way teaching right and wrong in their eyes. In a way creating a partial "Theory of Mind" AI. Preference toward conservative and liberal solutions can be guided by historical like kind solution grading. The exact same hardware with different historical event lessons will come up with different solutions. If only with the presentation of the solution. Just as humans do", says a netizen.*"

The opinions are divided over the issue.

"*The invention of Artificial Intelligence has definitely had a big and important impact on society. Some people say this impact is positive, but others think that the benefits it brings may cause addiction. In my opinion, people can use this advantage for the good of humanity and make their life easier than before. For example, when the first robots appeared some of the most difficult tasks could be solved efficiently and in short time. It is also interesting that some humanoid robots can interact with humans, making gestures or moving their heads. I like many of the uses where this new kind of intelligence can be applied. In the health area there are machines that can diagnose human illnesses and in the education area students can access additional help through computerized assistants. Machines can provide support when someone needs it, says another.*"

Others have expressed fears too.

"*The benefits that AI can bring to society will hopefully make the world better, but humanity must take control to assure the correct operation of the machines. I say this because I have read about machines getting smarter than people and developing their own conscience at levels that might harm human welfare. It is necessary to clearly establish limits to avoid misuse of the machine´s learning capacity that this new type of intelligence is reaching today.*"

By the year 2023, a self-regulating association for oversight on AI and machine learning developers may be established in some of the G7 countries. While regulating it can be a challenge, industries need to generate standardization around development as well as certification. This will help them work towards a set of professional standards for the ethical use of AI. Corporations should not neglect AI governance.

The companies have to be aware of potential regulatory as well as reputational risks. In fact, AI governance seeks to create policies in order to fight AI-related biases, discrimination, as well as other negative implications of AI. There are a few regulations governing the use of AI tools. Where laws do exist, they usually pertain only indirectly to AI.

For example, the Federal Fair Lending regulations need commercial and financial institutions to give details on credit decisions to potential customers. This, in a way, can restrict the degree to which creditors can use deep-learning algorithms. The GDPR puts severe limits on how companies can utilize consumer data. This can restrict the training and functionality of several consumer-facing AI applications.

Transparency of algorithms can reduce risks and boost confidence

In order to develop and cultivate AI governance, data analytics leaders need to focus on three areas: transparency, trust, and diversity. They will have to focus on trust and transparency in data sources as well as AI outcomes in order to guarantee successful AI adoption.

They also will have to recognize transparency requirements for data sources in addition to algorithms to decrease threats and help in boosting confidence in AI. They ought to warrant data, algorithms, as well as viewpoint diversity if they want to pursue AI ethics and accuracy. Some people believe that the term AI is too closely linked to pop culture, causing the public to have unfounded fears about it and giving rise to unrealistic expectations about how it will change work and life.

Researchers and marketers think that the name augmented intelligence, which has a more neutral connotation, may help people shed their fears and understand that this technology will simply improve products and services, and not replace the humans that use them. Do you agree?

The Age of AI

~~~~~~~~~~$$$$$$$$$$~~~~~~~~~~

CHAPTER SEVENTEEN

Race for AI Superpower Hots Up Among Nations of the World

Countries Race to Become AI Superpower

The current trends of Artificial Intelligence is sweeping industries across different countries. AI can recognise faces, run autonomous cars, deliver better outcomes and strengthen business. Not surprisingly, developed, developing, and under-developed countries are focusing on AI adoption for better opportunities and development. While AI is flourishing, major countries of the world are working towards winning the race. Let us look at what the leaders of the world are saying about AI.

"AI is the future and whoever becomes leader in AI will become the ruler of the world," says Russian president Vladimir Putin.

Chinese president Xi Jinping says: "China wants to be the world leader in AI by 2030".

"America has been the global leader in AI, and the Trump administration will ensure our great nation remains the global leader in AI", says US administration.

Similarly, there is a National strategy for Artificial Intelligence, published by NITI Aayog in India, whose vision is "AI-for-All in India".

The sayings signal the start of the struggle for attaining supremacy in the domain of Artificial Intelligence. In fact, it is already underway.

Is China the Next World Leader in AI?

China Leading the Race in AI

The country always had ambitions for becoming an AI Superpower of the world. Marching towards this goal, the State Council of People's

Republic of China has declared becoming a $150 billion AI global leader by the year 2030. The Chinese authorities are confident that AI is the future, thus building their infrastructure to be "future AI ready". For example, there is a new city being built near the capital Beijing, which has provisions for autonomous vehicles. It is building highways with sensors for driverless vehicles. The country has transformed itself and come up with a unique blend of entrepreneurism, innovation and value creation. Plus, the competition among Chinese entrepreneurs ensures that their products are at a level par excellence.

They excel at execution, speed of bringing products to the market, product quality, and better data utilization in decision-making. They are truly world leaders. The Chinese economy and market has a huge domestic testing ground and user base to try and test new concepts, products and services. The market is huge. It can absorb any product or service.

The dominance of the country is based on several factors. The goal looks attainable as China is already a global leader in AI research. China also focuses on AI Research. In the past few years, the country has produced the highest number of AI researcher papers and patents. Its population using the internet is approximately 750 million people, thus generating great amounts of digital data to process. This will, in all likelihood, pave its way to becoming AI superpower.

China has transformed itself into a Cashless, Card-Less, and Mobile-Only economy. This has created enormous amounts of useable data, useful for technology companies to accelerate AI development. Data is the fuel for AI. Chinese companies have digitized many services and have complete visibility due to the length, breadth and depth of data. Also, China has a powerful Venture Capital ecosystem, in which funds regularly flow for investments in companies and startups.

Besides, Chinese technologists and entrepreneurs are hungry for success. There is great backing from the Government on democratizing AI. On the other hand, protectionist measures provide local startups the bandwidth to establish and experiment in their early incubation period, that is, till the time they can compete with the big international players. Plus, Chinese education is more practical.

United States of America

UK in the AI Race

The country is giving good competition to China in the race for becoming AI superpower. There is an established technology culture in the US. About $10 billion venture capital has been channelled in the direction of AI. However, the funding has recently declined due to spike in education costs and immigration restrictions for global research professionals.

United Kingdom

In Europe, the UK is the undoubted leader with 121 AI-powered firms. Tech companies there were able to get an investment of $8.6 billion in the year 2017. This figure is 38 per cent of the entire venture capital investment that was done in the USA. Moreover, the United Kingdom government has announced funding of $78 mn for supporting robotics as well as AI research projects.

Canada

The government of Canada has been playing a key role as far as investments in AI projects are concerned. The Canadian government has committed $125 million investment in AI research. Importantly, after the United States presidential election was won by Donald Trump, Canada too started recruiting and spending on AI capabilities. So much so that the Quebec government has warned that it is essential to level ramp up investment and bring it to a level with that of the UK and China, otherwise the country may lag lack behind in becoming an AI world power.

Russia

Russian President Vladimir Putin is of the opinion that AI leader will rule the world. The country invests about $12.5 million annually on AI and research. Russia is focussing on AI in terms of participation of government in public as well as private AI engagements. A number of AI initiatives of Russia are military and strategic in nature, for instance AI-empowered fighter jets as well as automated artillery.

Germany

Germany

This nation is known for its efficiency and technical superiority. It is set to blend its tradition with that of technology innovations. Germany is marching towards a leadership position as far as robotics, autonomous vehicles, and quantum computing are concerned. Moreover, the Cyber Valley of Germany is attracting a lot of world-wide interest as well as investment.

Norway

Race for Artificial Intelligence Hots Up

This country has showcased its intent to expand it vision beyond its traditional resources such as oil drilling and fishing. In the race for technological advancements, it is ready to take the leap. Though, it has a long way to go in order to become an AI power, the country has put in place an accelerator program, which was launched in 2017 with $11 million of funding. This should indeed help Norway develop as a technological hub of the world and give some competition to the other contenders.

Sweden

According to a recent survey that was conducted in Sweden, 80 per cent of the country's residents are positive about AI in addition to robots. This means the country is all set to consider replacing human workers with automation and robots. They are ready to embrace AI and its technology and extend support to automation among various industries in Sweden. The Swedish union as well as workers have given green signal to AI as they feel it will enhance their human skills and give them a competitive edge in the global market.

France

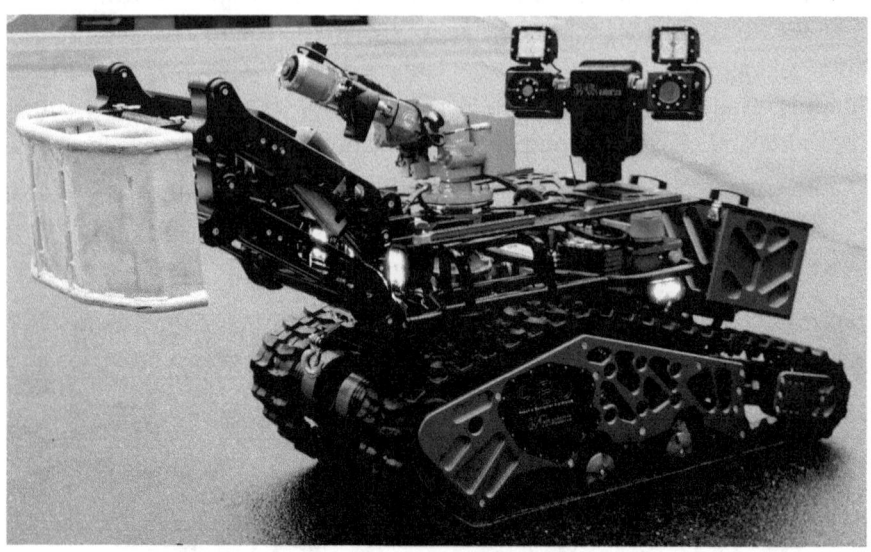

The government of France has pledged to invest $1.8 billion in AI research till the year 2022. The French AI initiatives are aimed at data strategy. Private companies need to publicly release data for fuelling the development of AI use cases. Other initiatives focus on research and how to strengthen it overtime. A specific amount of funding is likely to be invested in AI research partnership with Germany and other partners.

India: Country with Potential

Being a fast developing nation, India is going through transformation in digital space. It is one of the leaders in the field of IT. The impact of new technology such as AI on the country can be foreseen as the country seeks to cement its place as a technology leader in the world. Digital technologies are substantially contributing to the GDP. The percentage is likely to increase in the next few years. The country already has a huge number of IT professionals working with global companies and playing key roles in developing products and managing operations.

They understand products and business models and many of them have their own startups. Many of these are developing AI products in domains like healthcare, banking and finance, manufacturing, retail, agriculture, water management, and women empowerment, etc. The National strategy for Artificial Intelligence, India has been published by the NITI Aayog with the slogan "AI-for-All in India". The strategy identifies core areas where AI can play a crucial role.

AI can boost the Indian economy with an estimated annual growth rate of 1.3 per cent and add one trillion USD. India can provide a platform for enterprises to develop AI solutions to be implemented in the developing world and emerging economies. "Solved in India" mission is synced with

Artificial Intelligence as a Service (AIaaS). For example, Tata Memorial Hospital is working on the "Cancer Heat Map" to leverage AI for cancer detection and treatment.

Digital Pathology and Imaging Biobank can help in detection at an early stage. Indian start-up 'Forus Health' has developed a portable device "3Nethra" to screen for common eye problems like diabetic retinopathy. About 1.5 lakh Health and Wellness Centers and district hospitals will promote e-Health and AI is set to play a crucial role in its success.

In agriculture, around 50 Indian AI start-ups have been able to raise more than USD 500 million. For example, Intello Labs uses image-recognition technology and software to monitor crops and predict farm yields. Aibono leverages data science to stabilize crop yields. Trithi Robotics uses drone technology to monitor crops in real time. Crop health monitoring and real time action can increase utilization of farm machinery. AI-based tools can be leveraged for Soil Care, Sowing, Herbicide Optimization and Precision Farming. In Education, education systems are using adaptive learning tools for customized learning and interactive tutoring systems.

For example, Andhra Pradesh government is able to forecast School Dropouts using AI tools, which can also help in automated rationalization of teachers and development of customized professional courses. Investments has been made for building Smart Homes, Smart Parks, predictive service delivery, rationalization of administrative personnel, service demand and migration trend analysis, and grievance redressal. These solutions leverage some or other AI technology such as chat bots and smart assistants. Other examples are crowd management, safety systems, etc.

In Transportation and Smart Mobility, the AI technology can be leveraged in areas such as controlling congestion and road accidents. AI tools can help in areas such as Assisted Vehicles, Autonomous Trucks, Intelligent Transportation Systems, Travel Route Optimization and Community Based Parking. 'National AI Marketplace (NAIM)' and 'Data Marketplace' are also on the radar.

A hugely skilled resource pool of data scientists and AI engineers is already fertile ground. Plus, technical training in AI can cheap and accessible in India. There are forums to mentor young entrepreneurs and AI professionals. Many educational institutes follow the latest curriculum in data science and AI. Therefore, there is a conducive environment for AI to flourish as the country is already an IT powerhouse and can be the dark horse in the race of AI Superpowers.

~~~~~~~~~~$$$$$$$$$$$~~~~~~~~~~

www.ingramcontent.com/pod-product-compliance
Lightning Source LLC
Chambersburg PA
CBHW071524180526
45171CB00002B/378